HOW TO PASS

HIGHER CHEMISTRY

Martin Armitage

Hodder Gibson

A MEMBER OF THE HODDER HEADLINE GROUP

Acknowledgements

I should like to thank the following people whose help and support has been invaluable in the production of this book: my wife, Morag, for her constant encouragement; my daughter Lucy for her assistance in producing some of the illustrations; and my former colleagues Alison Boylés and Joe McKendrick as well as Angus Edward, who taught me most of what I know about teaching chemistry.

The Publishers would like to thank the following for permission to reproduce copyright material:

Photo credits

p.29 © Kevin Lamarque/Reuters/Corbis; **p.42** © Lew Robertson/Corbis; **p.43** © Nick Hawkes; Ecoscene/Corbis; **p.45** © Dex Images/Corbis; **p.51** © Charles E. Rotkin/Corbis; **p.54** © Chris Collins/Corbis; **p.112** © ImageState/Alamy.

Every effort has been made to trace all copyright holders, but if any have been inadvertently overlooked the Publishers will be pleased to make the necessary arrangements at the first opportunity.

Orders: please contact Bookpoint Ltd, 130 Milton Park, Abingdon, Oxon OX14 4SB. Telephone: (44) 01235 827720. Fax: (44) 01235 400454. Lines are open from 9.00–6.00, Monday to Saturday, with a 24-hour message answering service. Visit our website at www.hoddereducation.co.uk. Hodder Gibson can be contacted direct on: Tel: 0141 848 1609; Fax: 0141 889 6315; email: hoddergibson@hodder.co.uk

First published in 2005 by
Hodder Gibson, a member of the Hodder Headline Group
2a Christie Street
Paisley PA1 1NB

Impression number 10 9 8 7 6 5 4 3 2 1

Year 2010 2009 2008 2007 2006 2005

Cover photo shows sodium sulphate crystals © Science Photo Library
Typeset in 9.5 on 12.5pt Frutiger Light by Phoenix Photosetting, Chatham, Kent
Printed and bound in Great Britain by Arrowsmith, Bristol

A catalogue record for this title is available from the British Library

ISBN-10: 0-340-88792-3

ISBN-13: 978-0-340-88792-9

CONTENTS

INTRODUCTION

Congratulations – you've decided to try Higher Chemistry!

You enjoyed Standard Grade Chemistry and achieved a good result. You've picked a subject which will be a good qualification for Higher or Further Education. It will also give you a great understanding of the world around you, including environmental and industrial issues. As well as that, you'll learn skills useful in areas other than chemistry – 'transferable skills' like numeracy and problem solving.

The first thing to understand is that Higher Chemistry is a lot harder than Standard Grade. 20 000 students sit Standard Grade and half of them get Credit. About half of these Credit passes go on to Higher, but only about 20% of them get an A!

Standard Grade Chemistry contains very few calculations – about 4 or 5 marks worth out of 60 marks in the Credit exam. There are none in the General exam. At Higher, there are calculations worth up to 25% of the whole exam. This means that you need to get lots of practice at calculations in order to tackle them with confidence.

If you've bought this book, it's fair to assume that you already know quite a lot of chemistry. In it you'll find suggestions about how to revise. You'll find short summaries of each topic and examples of questions with discussions of how to tackle them. However, this isn't a textbook. It's a book about how to go about passing Higher Chemistry. The most important factors in passing are your approach to revision and the amount of work you do. This book isn't here to replace your teachers, your class notes or your textbook. It's here to help you to make the very best use of these.

Good luck!

Martin Armitage

Chapter 1

HOW TO REVISE

In Standard Grade, Knowledge and Understanding count for 40% of the marks. In Higher, it's 60%. You really have to know the course thoroughly. You just have to learn the work!

To make the point, let's try a little quiz. Can you answer these questions?

Questions

1. Name an element existing both as a covalent network and as discrete molecules.

2. Why is water (H_2O) a liquid, yet H_2S (a larger and heavier molecule) is a gas?

3. Why does every collision between molecules not result in reaction?

4. Name a fuel which causes no pollution of any kind.

5. What reagent converts an alcohol to an aldehyde?

You'll find the answers at the end of the chapter.

These are easy questions. They're pitched at level C – what you'd find in an NAB. Your answers should be almost automatic! If you don't know the answers to these, your Knowledge and Understanding isn't sound and you need to do something about it. You need to learn the basic facts and you need to start now.

What's in the Course?

So, what's in the course? What will you have to revise?

The course, like most National Qualifications courses, consists of three units:

Unit 1 ***Energy Matters***

Unit 2 ***The World of Carbon***

Unit 3 ***Chemical Reactions***

There's a test at the end of each unit (the dreaded NAB!). You need to pass all of them to get an award in Higher Chemistry at the end of the course – assuming you pass the final exam. NAB questions are much easier than the final exam questions, so don't be lulled into a false sense of security when you pass your first NAB – as you'll realise when you do your prelim.

Each unit is split into a number of topics – five or six in each unit. **Every** topic is tested in the final exam, so you must revise the whole course – there's no point trying to guess if something won't be tested (and therefore will not require revision) because it all will.

Getting Organised

Your course is probably taught topic by topic, with handouts, worksheets or notes for each. That means lots of paper by the end of the course. Get this mountain of paper under control. It's best to start by being organised, rather than trying to sort through a great heap of papers later on. Different methods suit different people, of course, but a cheap and easy way is to buy a box of transparent polythene pockets ('poly pockets') and to keep each handout in one of its own. Use a separate ring binder for each unit. That's three ring binders and some poly pockets.

Figure 1.1 Organisation is the name of the game

Date each handout as you receive it, so you can arrange them correctly in the ring binder.

It's a great idea to read over the notes the evening of the day you get them, to make sure that you can read them (it helps!), and that you know what they're about. If in doubt get help at your next visit to chemistry. It's best to do this when the lesson is fresh in your mind, as it makes learning a lot easier.

You might receive summary notes from time to time. But if you make your own, you'll learn your work a lot better!

When should I start?

Start revising when you start the course. OK, you have other homework, and it can be difficult to fit in additional revision. The least you should do is revise the **last** lesson in chemistry before the **next** lesson. That makes the new work much easier.

What should I revise?

Naturally, you need to revise all your **Higher** work! But some questions are based on **Standard Grade**. You can spend less time revising Standard Grade because a lot of Higher work consolidates Standard Grade. You get lots of practice at writing formulae and equations in Higher, as well as loads of organic structures and calculations so don't waste time revising

Figure 1.2 You can be asked loads of questions on the reactions of acids

these at Standard Grade. However, it's well worthwhile revising thoroughly the chemistry of acids, metals and neutralisation, as there is not too much of this in Higher, but lots of hard questions can be built on these topics.

How should I revise?

It's usually more efficient to revise 'actively'. Don't just sit and read over your notes. Practise writing down formulae, structures, summaries, equations, definitions, etc, and then check to see if you have done it properly.

'Flash cards' are a good way of doing some kinds of revision. If you want to remember the structure of ethanal, for example, cut out a piece of thin card and, on one side, write 'ethanal'. On the other, draw the ethanal structure.

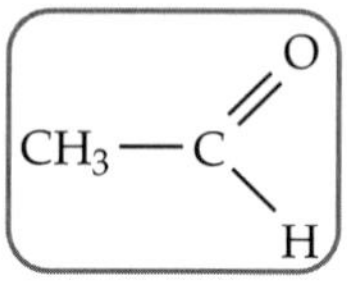

Test yourself by looking at the structure and naming the compound or vice versa. Flash cards are especially suitable for organic structures and functional groups in Unit 2. They're also good for definitions. For example, what's meant by:

- activation energy
- enthalpy of combustion
- Hess's Law
- weak acid
- enthalpy of solution.

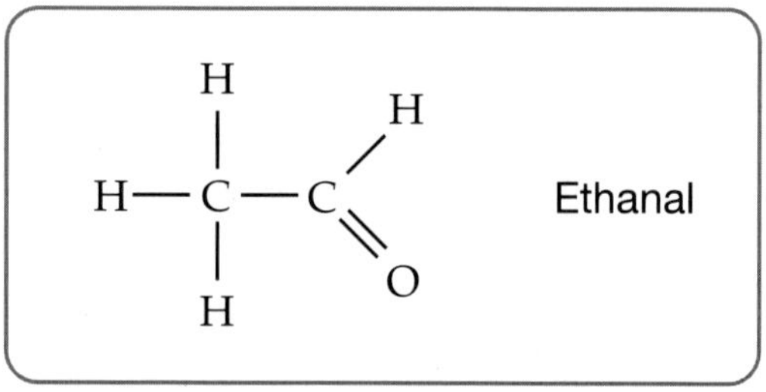

Figure 1.3 This side .... that side

Try making a **glossary** – a list of terms with their meanings. Then get someone else to help you by asking questions from the glossary. The great thing is that it needn't be someone who knows anything about chemistry! You'll find useful glossaries at the end of each unit on the 'Scholar' website, but making your own is better.

Mental maps and mnemonics

You can use 'mental maps' to make sense of chemical information. Pretend that what comes next is in your textbook or printed notes.

'Because all the atoms of an element are the same, there are only two possible kinds of bonding – metallic or pure covalent. Non-metallic elements all have covalent bonding, and exist either as discrete molecules or covalent networks.

If they exist as discrete molecules they will be gases or liquids at room temperature, or else solids with low melting points, because there is only van der Waal's bonding between the molecules. In the first 20 elements, the elements which form discrete molecules are the noble gases (monatomic), hydrogen, oxygen, nitrogen, and the halogens (fluorine, chlorine). Sulphur forms an S_8 molecule and phosphorus forms a P_4 molecule.

If they form covalent networks, they will be hard solids with high melting points. Only carbon, silicon and boron in the first 20 elements form covalent networks.'

Look at the 'mental map' below; see how it links the information.

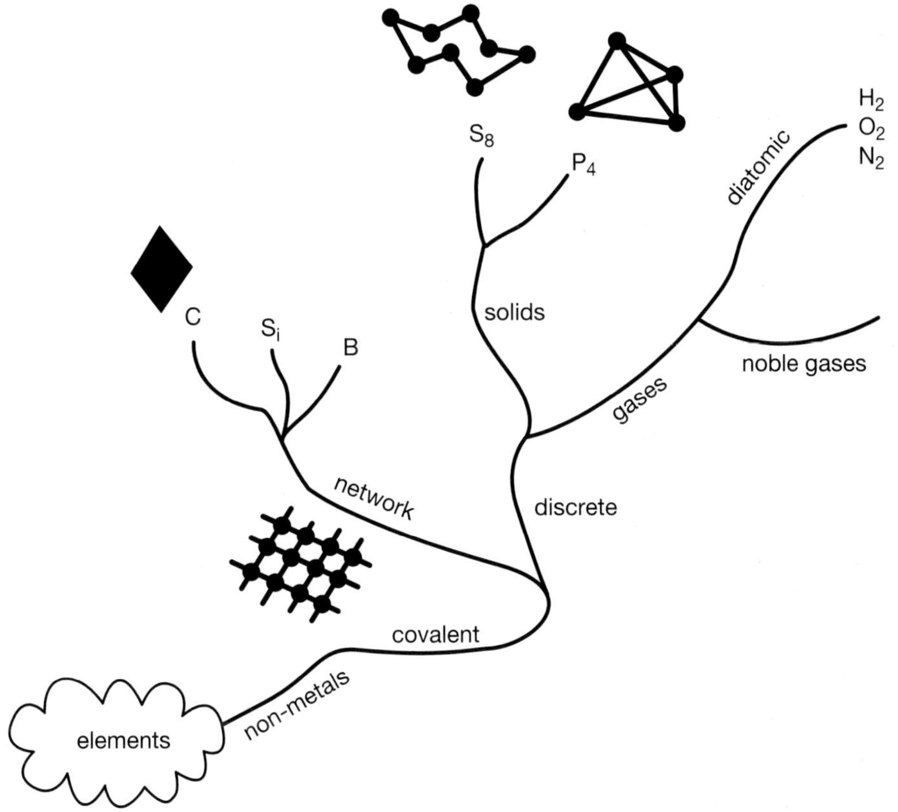

Figure 1.4 A 'mental map' for non-metals

It helps if you illustrate your 'map'. The example shows the shapes of the sulphur and phosphorus molecules, a diamond representing carbon in its covalent network form, and a 'network' to show that idea. There's no right or wrong way of making a mental map. It's how **you** see the information that matters.

A mnemonic is a way of making lists easier to remember. A mnemonic might take the first letters of each item in a list and use them to make a sentence. Each word in the sentence starts with the same first letter as the words in your list.

To help you remember the names of alkanes in the correct order, for example, you could make up sentences like '**m**y **e**lephant **p**lays **b**adminton' (to remind yourself of the correct order of **m**ethane, **e**thane, **p**ropane and **b**utane.)

Routine

Regular revision is best, and the sessions shouldn't be too long. Fifty minutes is a good length of time for efficient revision. You can break for a few minutes for coffee and start another 50 minute session of another subject. Quality is more important than quantity. Make yourself a timetable for revision. For example:

Monday		Tuesday		Wednesday		Thursday	
6p.m.	Chemistry	6p.m.	English	6p.m.	French	6p.m.	Biology
7p.m.	English	7p.m.	French	7p.m.	Biology	7p.m.	Chemistry
8.15p.m.	French	8.15p.m.	Biology	8.15p.m.	Chemistry	8.15p.m.	English
9.15p.m.	Biology	9.15p.m.	Chemistry	9.15p.m.	English	9.15p.m.	French

Remember, each slot is 50 minutes, so you have shorter and longer breaks. If you don't always like to follow biology with chemistry or some other combination, you can shuffle them more. The point is that it can suit some people always to follow the same routine. Others like to vary it a bit.

The example above shows four evenings. It's important to have a day off where you can do something quite different and be refreshed for the next set of revision days. Of course, you'll probably have some set homework as well, so you might not be able to work to this kind of routine until quite close to the actual exams.

Don't forget to take a break!

The Internet

There are some very useful websites which will help you with your revision.

Try these.

- www.scholar.hw.ac.uk
- www.evans2chemweb.co.uk
- www.bbc.co.uk/revision

There's also the SQA website.

- www.sqa.org.uk

Here, you'll find the most recent Higher exam and its marking scheme. You'll also find the 'Arrangements' for the exam. These tell you what's in your course, so you can check the progress of your revision. Print out the section of the arrangements on course content, and use it as a checklist for revision. It will also give you a measure of the pace of your revision. It's handy too, for making a note of things you don't understand, so you can ask your teacher or use one of the websites to get help.

What's the exam like?

Your exam will be for 100 marks.

Forty are for multiple-choice questions. In these you select one correct answer from four choices. If you pick **more** than one answer, then you'll get no mark for that question.

The answers will be divided roughly equally among As, Bs, Cs and Ds so don't expect all the answers to be A (or B or C or D for that matter!). Don't leave out a question if you don't know the answer. Any guess is better than no guess. You have one chance in four of being correct if you guess (better odds than the Lottery and a lot better than your chances of being struck by lightning).

Sixty of the marks are for 'extended response questions' – where you have to write the answer. Most of them need only pretty short answers – a word (or two) or a structure or formula. Look at the wording of the question and the number of marks for the question. If it says 'name the ....' or 'draw the structure of ......' that's all you have to do – name a compound or draw a structure.

If it says '**explain**' – for example '**explain** the trend in the first ionisation energies from lithium to neon' – you'll find the question is for two marks. To get them both you will first have to say what the trend is (....it increases...) then say why (... the nuclear charge increases). You'll get a mark for stating the trend, and a mark for giving a reason.

Figure 1.5 More chance of a multi choice mark than this happening to you!

An answer to an '**explain**' question has to go deeper than an answer to a '**state**' or '**name**' or '**write**' question.

A question may be for even more marks and you'll have to decide how much information to give in your answer.

The Data Booklet

Know your way around the booklet; it can help you in lots of ways. For example, on **page 6**, there is a table which tells you about the alkanes. They're given *in the order of increasing number of carbon atoms present*. The third one down is propane. That tells you that propane has three carbon atoms, and so on. You'll never have an excuse for getting the number of carbon atoms wrong in an organic compound.

On **page 6**, there's also a table which gives information on alcohols, aldehydes, ketones and acids. This is helpful. If your mind goes blank when you're trying to answer a question where you're to name such a compound, you'll find the names of all the common alcohols etc. there.

Don't forget that the booklet contains the formulae of ions containing more than one atom, with their charges so you shouldn't find it hard to write the formula for any compound, if required.

Page 8 contains two radioactive decay series. If you forget what happens in α or β emission, you can work it out from this table. This is because you can see how the mass number and the atomic number change during each type of emission.

RADIOACTIVE DECAY SERIES

Note In both tables γ emissions have been omitted.

TABLE 1 (Plutonium-Uranium)

Element	Symbol	Mass Number	Atomic Number	Type of Radiation	Half-life Period
plutonium	Pu	242	94	α	$3{\cdot}79 \times 10^5$ years
uranium	U	238	92	α	$4{\cdot}51 \times 10^9$ years
thorium	Th	234	90	β	24·1 days
protactinium	Pa	234	91	β	6·75 hours
uranium	U	234	92	α	$2{\cdot}47 \times 10^5$ years
thorium	Th	230	90	α	$8{\cdot}0 \times 10^4$ years

Figure 1.6 Part of the radioactive decay series from the Data Booklet

You are expected to know what is meant by the terms 'first ionisation energy', 'second ionisation energy' and so on. If you forget, look at the top of **page 10**; these terms are explained there. **Page 11**, the electrochemical series, will remind you of the charges of various ions, if you forget them.

The formulae of quite a number of chemicals (mainly acids) are given on **page 12**. This is a table intended for Advanced Higher candidates, but you could find it helpful if you forgot, say, the formula for **ethanoic acid**. It's there! (It's CH_3COOH.)

Answers to quiz

1. Carbon (as diamond, a covalent network, and fullerene, discrete molecules)

2. Water has hydrogen bonding, which holds the molecules together. Hydrogen sulphide has no hydrogen bonding.

3. Because not every collision involves molecules with the energy of activation.

4. Hydrogen – when it burns, it produces water only.

5. Acidified potassium dichromate or hot copper oxide.

ENERGY MATTERS

Unit 1 is 'Energy Matters'. It covers the basic rules of how atoms and molecules behave, energy changes and the mole.

The Unit comprises the following topics.

a) Reaction rates
b) Enthalpy
c) Patterns in the Periodic Table
d) Bonding, structure and properties
e) The mole

Reaction rates

Summary

You already know that the rate of a reaction is affected by the temperature, concentration and the size of solid particles and that catalysts and enzymes speed up reactions. At Higher, you go a lot deeper into the theory.

Everything is made of constantly moving molecules. Low energy molecules move slowly, high energy molecules, quickly. Most move with 'average' speed – they have average energy. When molecules collide, they may react, providing they've got enough energy (Activation Energy, E_A). Only the fastest have this. On collision, such molecules may form an 'activated complex' which can break down to form products. At higher temperatures, molecules move faster. More of them have the Activation Energy, so more of them react.

In Figure 2.1 T_2 is a higher temperature than T_1. At T_2 many more molecules reach E_A.

Increasing concentration also causes more collisions, but molecules still require the Activation Energy.

Following the course of a reaction

Consider this reaction being done in a beaker:

$$CaCO_3 + 2HCl \longrightarrow CaCl_2 + CO_2(g) + H_2O$$

You could place the beaker on a balance and follow the mass loss as gas is given off. You could use a gas syringe to measure the gas volume as time passes. You could measure pH as time passes, since acid is used up. You calculate rate by dividing the change in, say, mass by the time taken for the change.

Catalysts

Catalysts speed up reactions. Homogeneous catalysts are in the same state as the reactants; heterogeneous catalysts are in a different state. Catalysts let a different, lower energy, activated complex form, still giving the same products. Heterogeneous catalysts adsorb reactant molecules onto active surface sites in an arrangement which lets an activated complex form with the other reactant. Catalytic poisons bond so firmly to the active sites that there are no sites for reactant molecules.

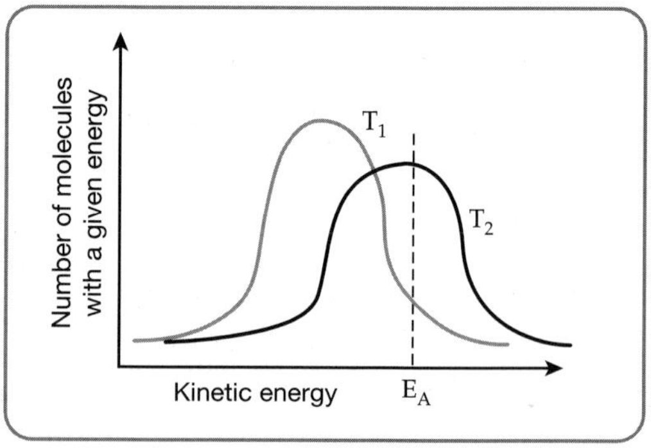

Figure 2.1 Energy distribution of molecules at two different temperatures

Example 1

The reaction between propanone and iodine is catalysed by a solution of hydrogen ions.

$$CH_3COCH_3(aq) + I_2(aq) \longrightarrow CH_3COCH_2I(aq) + HI(aq)$$

a) Why can you describe the catalyst as **homogeneous**?

b) If the reaction is carried out at a higher temperature, it is faster.
Use an energy distribution diagram to explain this.

Solution

a) You see that the reactants – the chemicals on the left of the arrow – are present as aqueous solutions. So is the catalyst. They are in the same physical state, and the catalyst is therefore homogeneous.

b) You can simply use the diagram shown in Figure 2.1.

Example 2

Two experiments were carried out to study the reaction of zinc with hydrochloric acid in an open beaker.

The result of Experiment 1 is shown in Figure 2.2.

a) Why did the balance record a decrease in mass during the reaction?

b) The only difference between Experiment 2 and Experiment 1 was the use of a catalyst. On the above graph, add a curve which you would expect for Experiment 2.

Solution

a) There was a mass loss because a gas was given off. The equation for the reaction is:

$$Zn + 2HCl \longrightarrow ZnCl_2 + H_2.$$

Note that this question requires a knowledge of Standard Grade – the reactions of metals and acids.

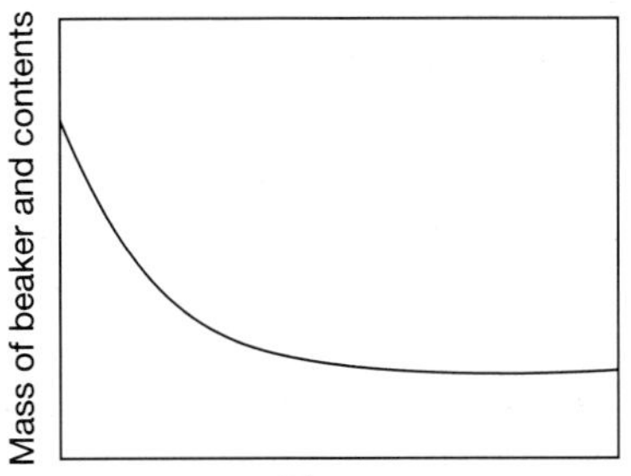

Figure 2.2 Graph showing change in mass with time

b) The catalyst speeds up the reaction, so the mass falls faster than in the case of Experiment 1. However, the catalyst doesn't result in you getting more product – you just get it faster. As a result, the curve will fall more steeply than the given curve but level off at exactly the same mass.

The required curve is shown by the dotted line.

Such a question might involve the award of a $\frac{1}{2}$ mark for the steeper slope, and another $\frac{1}{2}$ for flattening out at the same level as the original curve.

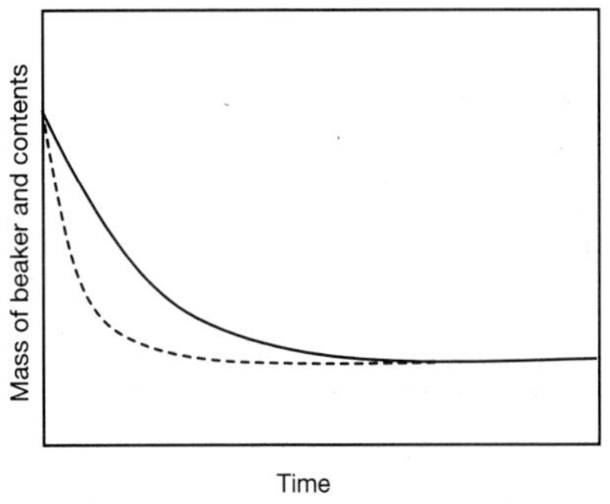

Figure 2.3 Graph showing change in mass with time

Example 3

Which of the following is **not** a true statement about the operation of a catalyst?

A A catalyst provides an alternative reaction pathway.

B A catalyst lowers the Activation Energy.

C A catalyst provides energy so that more molecules have successful collisions.

D A catalyst forms bonds with molecules of reactant.

Solution

Remember, a catalyst forms an alternative activated complex, which breaks down to give the same products. That means A is true. The alternative activated complex requires a lower Activation Energy, so that means B is true as well. If the catalyst forms an activated complex, it must form bonds with the reactants, so D must be true. This leaves C as the untrue statement. The answer is **C**.

Enthalpy

Summary

Enthalpy is heat energy, with units of kilojoules per mole ($kJ\,mol^{-1}$).

When heat's given out, the reaction is exothermic. If it's taken in, it's endothermic. Enthalpy changes are shown using the symbol ΔH, negative in exothermic reactions, positive in endothermic.

The enthalpy changes in a reaction are often shown in diagrams like these.

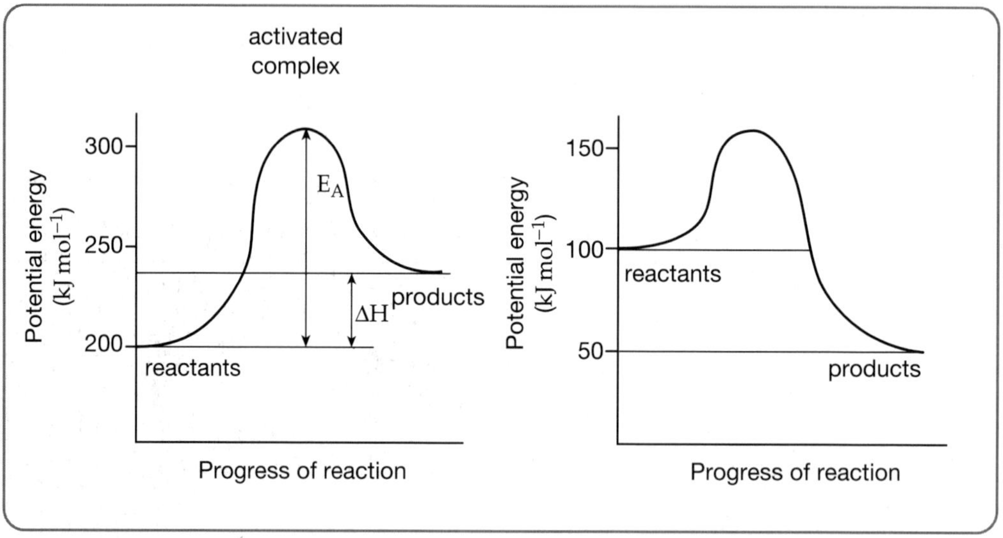

Figure 2.4 Potential energy diagrams for endothermic and exothermic reactions

The left-hand diagram shows products with a higher energy than the reactants – an endothermic reaction. The right-hand diagram shows an exothermic reaction.

The energy gap between the reactants and the activated complex is the Activation Energy for the reaction.

The energy difference between reactants and products is the Enthalpy of Reaction, shown by ΔH. In the left-hand diagram, it's positive – an endothermic reaction. In the right-hand diagram, it's negative – an exothermic reaction.

In the left-hand diagram, the product enthalpy is $240\,kJ\,mol^{-1}$

The reactant enthalpy is $200\,kJ\,mol^{-1}$. The enthalpy has increased by $40\,kJ\,mol^{-1}$ so ΔH is $+40\,kJ\,mol^{-1}$.

The Activation Energy is $100\,kJ\,mol^{-1}$.

You don't need a sign before the Activation Energy – it's always positive.

Example 4

Look at the potential energy diagram.

The energy of activation (E_A) for the forward reaction is given by:

A R

B Q – P

C R – P

D R – Q

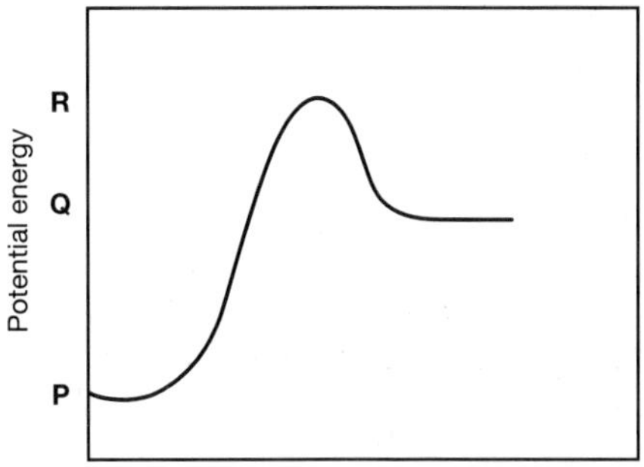

Figure 2.5 Potential energy diagram

Solution

You should know that the answer to this is simply the energy difference between reactants and activated complex. It must be R – P. The answer, therefore, is **C**.

In fact, R is the energy of the activated complex.

Q – P is the enthalpy change, ΔH, for the reaction.

R – Q is the energy released when the activated complex turns into product.

(It's also the activation energy for the reverse reaction.)

Example 5

The energy changes for this reaction

$$CO + NO_2 \rightarrow CO_2 + NO$$

are shown in the diagram.

ΔH, in $kJ\,mol^{-1}$, for this reaction is:

A –361

B –227

C –93

D +361

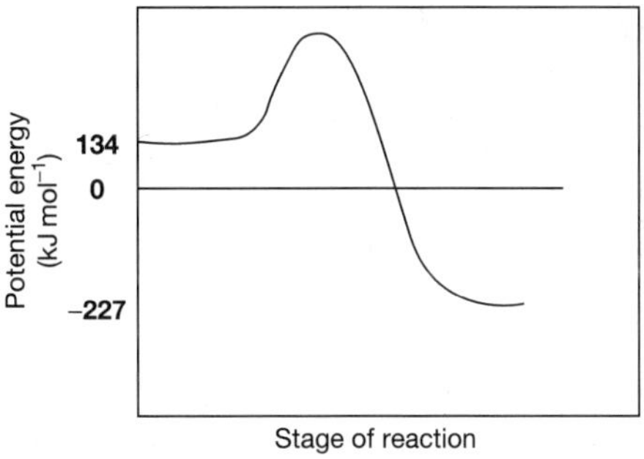

Figure 2.6 Potential energy diagram

Solution

Forget answer D. That would be an endothermic reaction, with a higher product energy than reactant energy. This is obviously an exothermic reaction, since the products have less energy than reactant energy.

You should spot that the energy drops from +134 to zero and then drops further to –227. This is a total drop of 134 + 227 giving a total drop of 361. Since it's a drop in energy, $\Delta H = -361\ kJ\,mol^{-1}$. The answer is therefore **A**.

Enthalpy and catalysts

Catalysts lower the activation energy for a reaction, by providing an alternative activated complex.

You may sometimes see questions like this:

Question

Here is the energy diagram for a reaction.

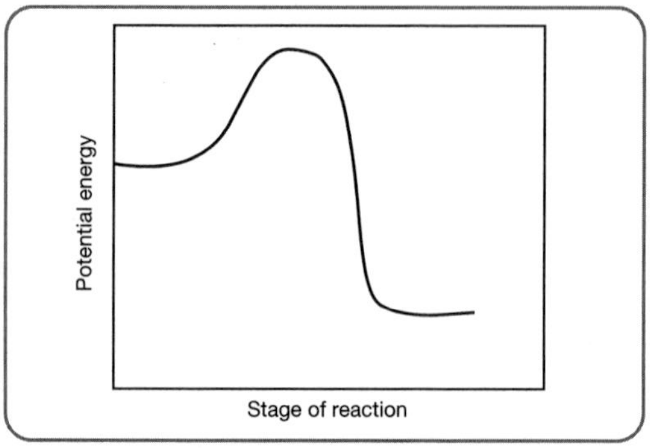

Figure 2.7 Potential energy diagram for a reaction

Add a dotted line to show the effect of adding a catalyst.

A catalyst lowers the activation energy, so you should draw the line as follows.

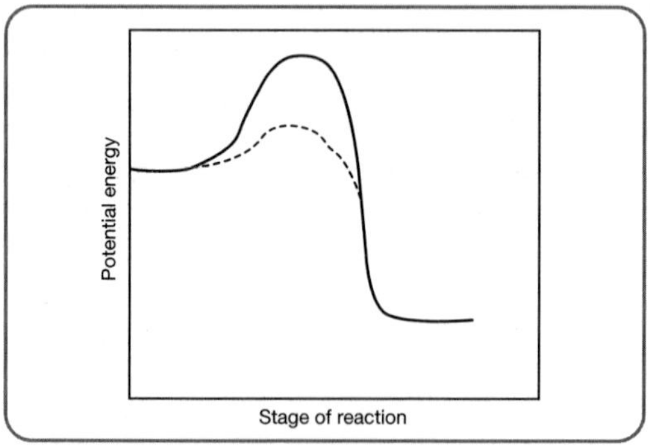

Figure 2.8 Potential energy diagram showing effect of using catalyst (dotted line)

The catalyst doesn't affect the energy of reactants or products, so ΔH stays the same.

Example 6

Which of the following describes the effect of a catalyst on a reaction?

	Enthalpy of reaction	Activation energy
A	Increased	Decreased
B	Decreased	Unchanged
C	Decreased	Decreased
D	Unchanged	Decreased

Solution

If you've understood what it says above, you should have no hesitation in picking **D** – no effect on enthalpy of reaction, and a decreased Activation Energy.

Other enthalpy changes

We'll deal with these in the 'Calculations' chapter.

Patterns in the Periodic Table

Summary

You need to know the layout of the Periodic Table, and how the physical properties of elements change from region to region of the table.

Elements in the same group have the same number of outer electrons so they have similar chemical properties. As you descend the group, more electron shells are added.

As you go along a period, an additional proton is added to the nucleus each time.

Metals are found at the left-hand side of the table, and non-metals on the right.

Variation in size (covalent radius)

As you go along a period, atoms get smaller. Increasing numbers of protons in the nucleus attract the electrons more strongly, pulling them inward. As you go down a group, the atoms get bigger, because there are more electron shells. Density also increases down a group.

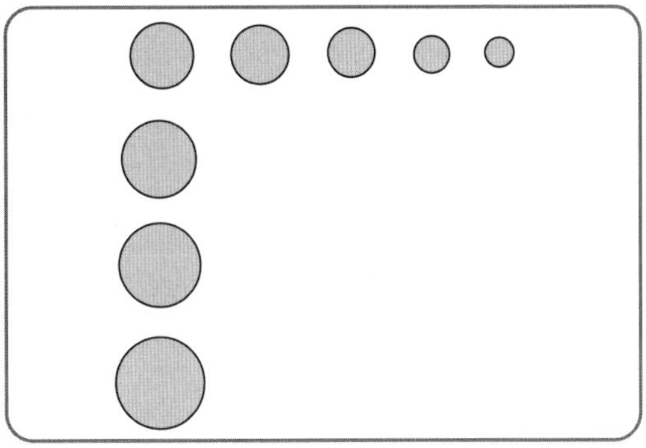

Figure 2.9 Trends in atom sizes across a period and down a group

Ionisation energy

The First Ionisation Energy of an element is the energy, in $kJ\,mol^{-1}$, needed to remove one mole of electrons from one mole of the atoms.

The equation for this change is $E(g) \rightarrow E^{+}(g) + e^{-}$

The second ionisation energy is $E^{+}(g) \rightarrow E^{2+}(g) + e^{-}$

Note the state symbols – very important here!

A chemical joke!

A sodium atom said to its friend: "I've lost an electron". "Are you sure?" said the friend. "Yes, I'm positive!" said the sodium atom. Think about it!

Example 7

Calculate the energy required for the following process.

$$Mg(g) \rightarrow Mg^{2+}(g) + 2e^{-}$$

Solution

This is a tricky one. You might think that you're looking at the second ionisation energy of magnesium. But think about it. First you have to remove one electron. Then you have to remove a second. What you really want is the sum of the first and second ionisation energies. Your answer is 744 + 1460 = **2204 kJ mol^{-1}**.

In a similar way, the energy required for

$$Al(g) \rightarrow Al^{3+}(g) + 3e^{-}$$

would be the sum of the first three ionisation energies.

Trends in ionisation energy

Ionisation energy increases along a period because the increasing nuclear charge makes it harder to remove electrons. It decreases down a group, since the outer electrons are further from the nucleus and are 'screened' from nuclear attraction by more inner electron shells.

Electronegativity

Electronegativity is a measure of the ability of atoms to attract electrons in a bond. Elements at the top right of the table (O, N, F and Cl) are very electronegative.

Elements at the bottom left are the least electronegative.

Example 8

Identify the correct trend as the relative atomic mass of the halogen increases.

A The atomic radius decreases.

B The density decreases.

C The ionisation energy decreases.

D The boiling point decreases.

Solution

You could check all of this out using suitable pages of the Data Booklet, but it should be clear that if the relative atomic mass increases, so must the atomic radius, the density and the boiling point, because the atoms are getting bigger and they're all in the same group. That leaves only C. The ionisation energy decreases as the atoms get bigger because the outer electron is further from the nucleus, and because of the 'screening' effect.

Example 9

Which equation represents the first ionisation energy of a diatomic element, M_2?

A $\frac{1}{2}M_2(s) \rightarrow M^+(g)$

B $\frac{1}{2}M_2(g) \rightarrow M^+(g)$

C $M(g) \rightarrow M^+(g)$

D $M(s) \rightarrow M^-(g)$

Solution

This is a bit harder, because the electrons have been left out! But if you remember that the element has to be in the form of separate gaseous atoms you have to rule out A and D, since the element is solid, and B, because the element hasn't been converted to separate atoms. Besides, you don't get negative ions when you lose electrons. Remember the sodium atom! C is the answer. It doesn't matter if the element is diatomic – ionisation energy refers to individual atoms.

Example 10

The Data Booklet contains the following information:

	Ionisation Energies/kJ mol^{-1}	
Element	First	Second
hydrogen	1311	
helium	2380	5260

a) Why are there two ionisation energies for helium, but only one for hydrogen?

b) Why is helium's first ionisation energy higher than hydrogen's?

Solution

a) The answer to this is simple. Hydrogen has an atomic number of 1. It only has one electron to lose – it can't lose a second.

b) Helium has two protons in its nucleus. They will have a stronger attraction for electrons than the one proton of hydrogen. It's harder to remove an electron, so the ionisation energy is higher.

Bonding, Structure and Properties

Summary

You can use electronegativity to make predictions about the kind of bonding between atoms.

- Atoms with very different electronegativities tend to form ionic bonds.
- Atoms with very similar or identical electronegativities form covalent bonds.
- Atoms with 'fairly' different electronegativities form polar bonds.

Bonding in elements

There are two kinds of element – metal and non-metal.

Metals have metallic bonding. The outer electrons ('delocalised') are free to leave the metal atoms, and flow round the resulting ions. The diagram gives an idea of this. The shaded grey area represents the delocalised electrons.

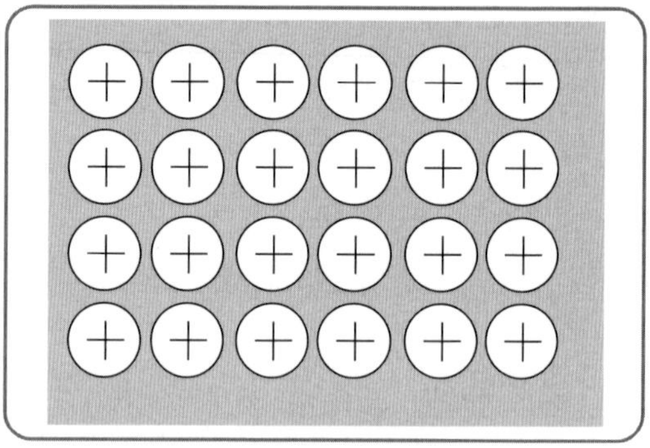

Figure 2.10 Metallic bonding

Non-metals have pure covalent bonding – the atoms have identical electronegativities.

Bonding in compounds

If elements are far apart in the Periodic Table, there is a big difference in their electronegativities. The bonding will be ionic. If they are closer together, their electronegativities are similar and the bonding will be covalent.

Example 11

In which molecule will the chlorine atom carry a partial positive charge?

A Cl – Br

B Cl – Cl

C Cl – F

D Cl – I

Solution

In each molecule, the atom with the higher electronegativity will carry a slight negative charge and the atom with the lower electronegativity will carry a positive charge. It's just a case of comparing numbers. The electronegativities are as follows.

Fluorine	4·0
Chlorine	3·5
Bromine	2·8
Iodine	2·6

The answer is **C**. Fluorine is more electronegative than chlorine so in this case the chlorine has the partial positive charge.

Example 12

Which compound is most likely to have ionic bonding?

A Beryllium chloride

B Aluminium chloride

C Potassium fluoride

D Nitrogen bromide

Solution

If you use the Data Booklet to find the electronegativities of each element in the compound and calculate the difference you will find:

A 1·5

B 1·5

C 3·2

D 0·2

The biggest difference is in C. Potassium fluoride is probably ionic.

Of course, you could simply note that in this question potassium and fluorine are the pair of elements furthest apart in the Periodic Table and assume therefore that the compound is ionic.

If the question had asked you which compound was most likely to have covalent bonding, you would have picked D, because the electronegativity difference is least.

Despite the fact that A and B both consist of a metal and a non-metal, the bonding is, in fact, polar covalent in each compound. Don't assume that the combination of a metal with a non-metal automatically leads to ionic bonding.

Example 13

What kind of bonding would be expected in each of the following?

a) Potassium fluoride

b) Chlorine

c) Carbon monoxide

d) Magnesium

Solution

a) There is a large difference in electronegativity – ionic.

b) Chlorine is Cl_2. There are two identical atoms – pure covalent.

c) There is not a large difference in electronegativity – polar covalent.

d) It's a metal – metallic bonding!

Structure in elements

You're expected to know the structures of elements 1 to 20. Metals have a metallic structure with a lattice of positive ions held in position by delocalised electrons.

Non-metals have a variety of structures, but the bonding is always pure covalent.

The following form diatomic molecules: hydrogen, oxygen and nitrogen as well as the halogens. This sentence might help you: *'I bring clay for our new home'*. Why?

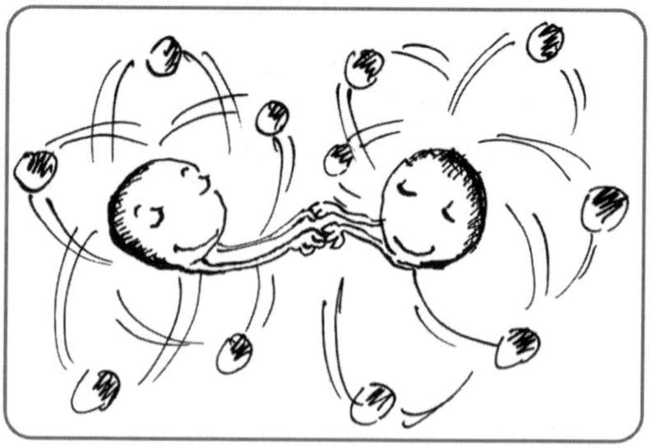

Figure 2.11 In diatomic molecules the atoms pair up

The noble gases are monatomic (single atoms), while sulphur and phosphorus form discrete covalent molecules, S_8 and P_4. 'Discrete' just means 'separate'.

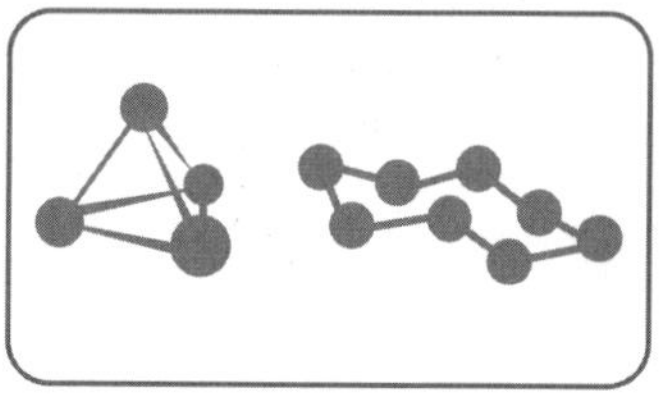

Figure 2.12 P_4 and S_8

Carbon forms covalent networks – diamond and graphite. The diamond structure involves carbon atoms with four bonds to other carbon atoms, based on a tetrahedral structure. Graphite has layers of hexagons which can slip over each other. Silicon and boron also form covalent networks.

Carbon also forms fullerenes, which are discrete molecules.

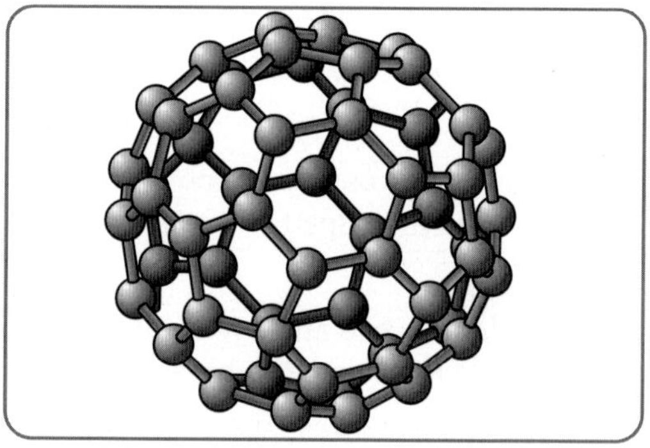

Figure 2.13 A fullerene

Structure in compounds

If the bonding is ionic, then the structure will be an ionic lattice, with alternating positive and negative ions. If the bonding is covalent, the compound will exist as discrete molecules.

Silicon dioxide and silicon carbide form covalent networks.

Polar molecules

Polar bonds may make the molecule, as a whole, polar.

An oxygen chloride molecule has this arrangement of atoms:

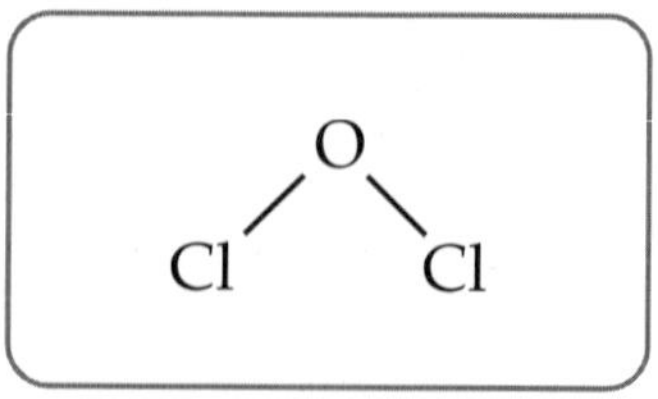

Figure 2.14 Oxygen chloride molecule

Oxygen is more electronegative than chlorine, making the oxygen end of the molecule slightly negative compared to the chlorine end. This makes the whole molecule polar.

A carbon chloride molecule has this arrangement of atoms.

This is more symmetrical than the OCl_2 molecule. All the bonds are polar due to electronegativity differences. The 'symmetrically opposed polar bonds' make the bond polarities cancel each other out so the molecule as a whole isn't polar

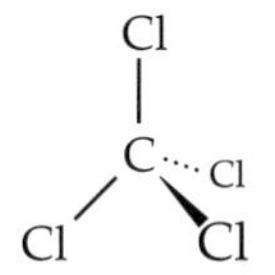

Figure 2.15 Carbon chloride molecule

Polar molecules attract each other. This increases the boiling point of the compound.

Example 14

Here are the shapes of some common molecules. They all contain polar bonds.

Which molecule below is non-polar?

A

B

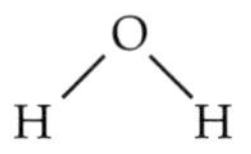

C

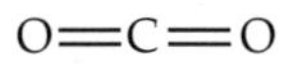

D

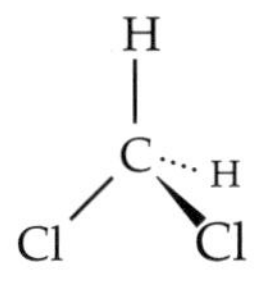

Solution

This question is to do with the symmetry of the molecules. Which molecule is so symmetrical that the polar bonds will cancel each other?

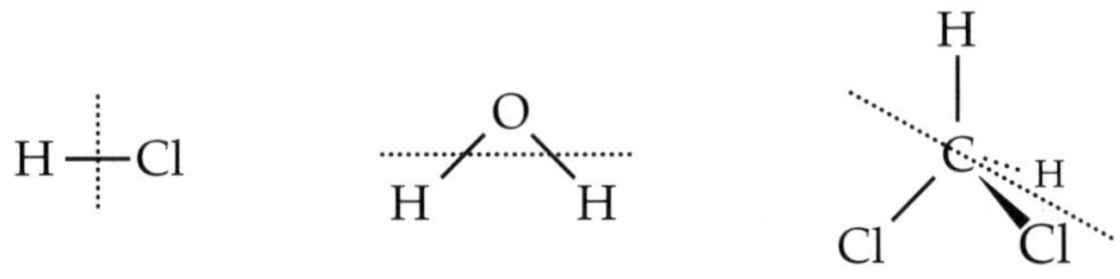

Figure 2.16 Polarised molecules

Look at C, the carbon dioxide. It's a linear molecule, with two polar C = O bonds directly opposite each other. The effect of this is to cancel the polarities. In the three other molecules, one end looks different from the other.

Example 15

Which of the following compounds will consist of polar molecules?

A CH_4

B CO_2

C NH_3

D N_2

Solution

You can rule out D right away. It's an element. Both atoms have the same electronegativity so there are no polar bonds. In that case, the molecule can't be polar.

You're expected to know that methane (CH_4) and carbon dioxide (CO_2) are highly symmetrical. But think about ammonia (NH_3). It has this shape.

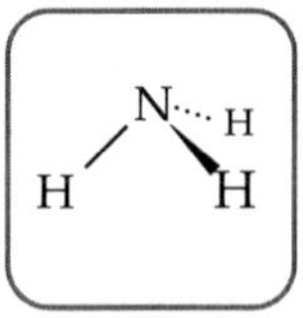

Figure 2.17 Ammonia molecule

It's fairly symmetrical – but the nitrogen atom at the top of the pyramid is much more electronegative than the three hydrogen atoms. The molecule as a whole is polar. The top of the pyramid is negative and the three hydrogen atoms are positive.

Hydrogen Bonding

The most polar bonds are formed between hydrogen and the elements fluorine, nitrogen and oxygen. When molecules contain these bonds, there are very strong attractions between the molecules. This is called 'hydrogen bonding'. It raises the boiling point substantially. Water, H_2O, has a formula mass of 18. Any other covalent substance with a similar formula mass (e.g. methane) is a gas, yet water is a liquid.

Example 16

When ice melts, which kind of bond is broken?

A Covalent

B Ionic

C Metallic

D Hydrogen

Solution

It shouldn't be too hard to pick the right answer! When you melt ice, you separate the water molecules from each other – you don't break the actual molecules – to do that, you'd have to break the covalent bonds holding the hydrogen and the oxygen together. The ice is held together by hydrogen bonds. The answer is D.

Van der Waals forces

All molecules attract each other weakly by Van der Waals forces. These forces result from molecules becoming temporarily polar due to a momentary lack of symmetry in the way their electrons are arranged.

Intermolecular bonding and intramolecular bonding

Intermolecular bonding means bonding **between** molecules, for example, hydrogen bonding, van der Waals forces and attractions between polar molecules.

Intramolecular bonding is bonding **within** a molecule i.e. covalent bonding.

Example 17

Which type of bonding can be described as intermolecular?

A Covalent bonding

B Hydrogen bonding

C Ionic bonding

D Metallic bonding

Solution

Easy, isn't it? **B** – hydrogen bonding. Ionic compounds and metals aren't even made of molecules! And covalent bonds are found **in** molecules – intramolecular.

Properties of ionic and covalent compounds

Different kinds of bonding result in different properties. Here's a summary of some of the main differences.

Ionic compounds	Covalent compounds
Solid at room temperature	Usually liquid or gas at room temperature.
Conduct electricity when melted or dissolved	Do not conduct electricity in any state
Often soluble in water	In general, not soluble in water

The difference in state at room temperature results from the strong ionic bonds in ionic compounds, and the weak van der Waals forces in covalent compounds. The presence of charged particles with the ability to move makes ionic compounds conduct when liquid. Ionic compounds are often water soluble because of attraction between the ions and polar water molecules.

Example 18

Explain why carbon dioxide is a gas at room temperature while silicon dioxide is a solid.

Solution

This question draws on quite a lot of knowledge. You're expected to know that silicon dioxide forms a covalent network, which is a very strong structure in which all the atoms are interconnected to all the others. Carbon dioxide is a linear molecule and although the individual $C = O$ bonds are popular, the molecule is very symmetrical and the polarity is cancelled out. The only forces of attraction between the molecules are the very weak van der Waals forces. So carbon dioxide is a gas. There's nothing to hold the individual molecules together.

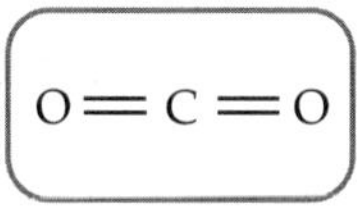

Figure 2.18 Carbon dioxide molecule

Example 19

Complete the following table.

Element	Melting point/K	Bonds or forces broken at the melting point
Sodium	371	
Silicon	1683	
Phosphorus	317	

Solution

The answer to 'sodium' should be clear. Metals contain metallic bonds.

You're expected to know that silicon forms a covalent network. Each atom is interconnected to all the others by covalent bonds. The only way you can melt this is by breaking the covalent bonds. These bonds are strong – just look at the temperature required!

You're also expected to know that phosphorus forms discrete molecules. The formula is P_4. Because all the atoms are the same, the molecules can't be polar. All that holds them together are the weak van der Waals forces. That accounts for the low temperature required to melt phosphorus.

Example 20

Propanone (CH_3COCH_3) and butane (C_4H_{10}) have the same relative formula mass of 58.

Explain why butane is a gas at room temperature (BP –1°C), while propanone is a liquid (BP 56°C).

Solution

Look at the structures of propanone and butane.

In propanone, the $C = O$ bond is polar, since the electronegativity of carbon is 2·5 and that of oxygen is 3·5. The bond is polar with oxygen negative and carbon positive. Because this polar bond is not cancelled out by one on the other side of the molecule, the molecule as a whole is polar.

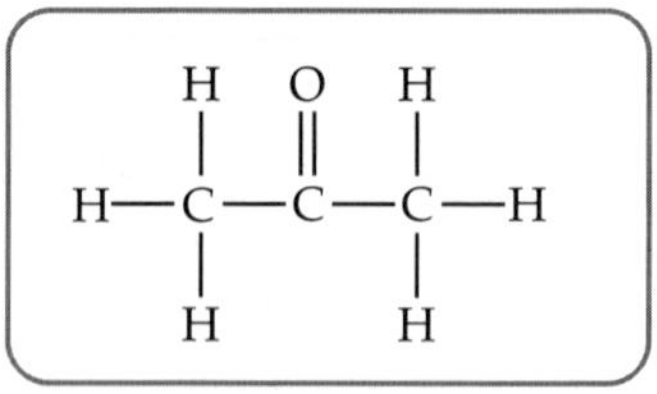

Figure 2.19 Structure of propanone

Butane is a symmetrical molecule, so any polar bonds are balanced out and cancelled.

Since propanone is polar, its molecules attract one another and are harder to separate than those in butane. As a result, propanone is a liquid and butane is a gas.

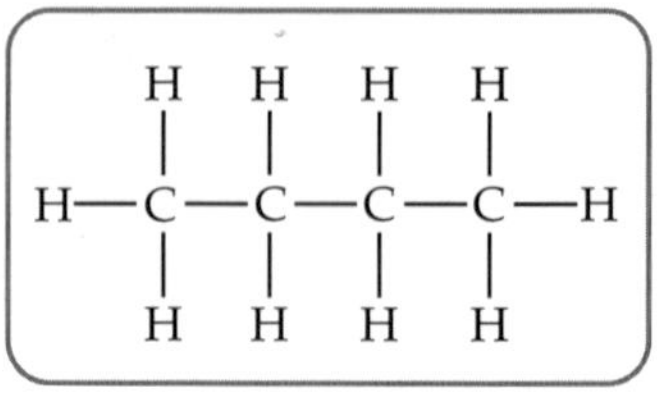

Figure 2.20 Structure of butane

Example 21

Glycerol is a very viscous liquid. It has this structure.

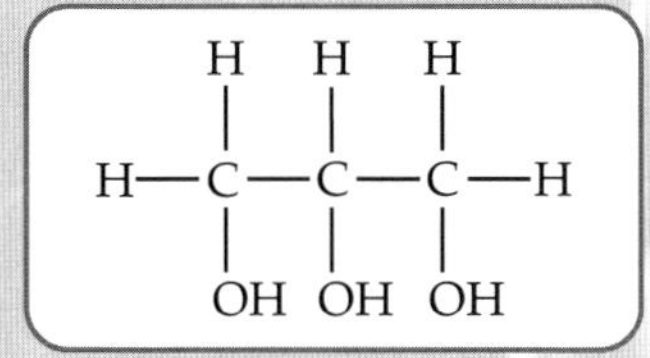

However, propane (C_3H_8) is a gas at room temperature. Glycerol is soluble in water, while propane is not.

Explain these differences in properties.

Solution

As you can see, the glycerol has three –OH groups on it. The presence of these will lead to a huge amount of hydrogen bonding between glycerol molecules. This causes the molecules to clump together, accounting for glycerol's viscosity or thickness. The only bonding between propane molecules results from van der Waals forces. These are very weak, so there is little to hold the molecules close to each other.

Both water and glycerol contain –OH groups. This means that hydrogen bonding can also take place between water molecules and glycerol molecules. This results in glycerol being soluble in water.

The Mole

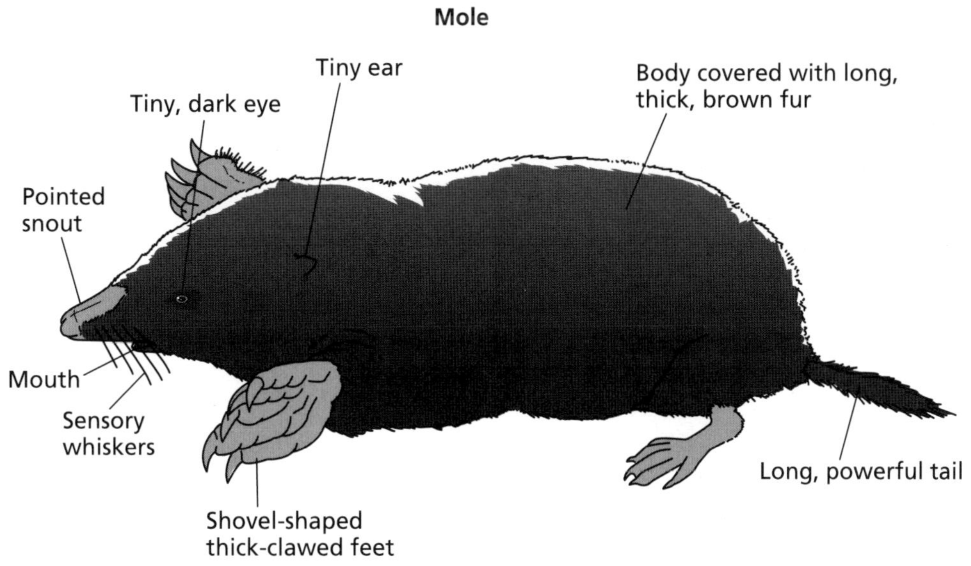

Figure 2.21 But not this kind!

Summary

In Standard Grade a mole of a substance was defined to be the Relative Formula Mass expressed in grams. So a mole of water, H_2O, was 18 g.

At Higher, you've learned that a mole of a substance contains $6{\cdot}02 \times 10^{23}$ particles of the substance.

Molar volume is the volume occupied by one mole of gas at a particular temperature and pressure. A mole of any gas, at room temperature and a pressure of one atmosphere, occupies about 24 litres. The more moles of gas, the bigger the volume it occupies. It follows that the volumes of gases are directly proportional to the number of moles.

Example 22

Which of the following contains the most molecules?

A 0·10 g of hydrogen

B 0·17 g of ammonia

C 0·32 g of methane

D 0·35 g of chlorine

Solution

To tackle this, you need to know the formula of each compound.

These are:	hydrogen	H_2 (diatomic!)
	ammonia	NH_3
	methane	CH_4
	chlorine	Cl_2 (diatomic again)

Then you need to work out the mass of 1 mole of each.

H_2	2 g
NH_3	17 g
CH_4	16 g
Cl_2	71 g

Now you calculate the number of moles of each compound. You divide the given mass by the mass of 1 mole and find:

H_2	0·05 mole
NH_3	0·01 mole
CH_4	0·02 mole
Cl_2	0·005 mole

The number of molecules is proportional to the number of moles. So hydrogen wins. **A** is the answer.

Example 23

The density of chlorine gas is found to be 3·00 g ℓ^{-1}.

Use this information to calculate the molar volume of the gas.

Solution

Chlorine's relative atomic mass is 35·5. Since chlorine is diatomic, Cl_2, one mole of chlorine has a mass of 71 g. If 3 g of chlorine occupy 1 mole, then 71 g will occupy 71 ÷ 3 = **23·7 litres**.

You can see that, if you know the density of a gas you can calculate its molar volume. This means that if you know its molar volume, you can calculate the density.

Example 24

Diphosphine, P_2H_4, is a hydride of phosphorus. All of the covalent bonds in diphosphine are non-polar because the elements present have the same electronegativity.

a) What is meant by the term 'electronegativity'?

b) The balanced equation for the combustion for diphosphine is:

$$2P_2H_4(g) + 7O_2(g) \longrightarrow P_4O_{10}(s) + 4H_2O(l)$$

What volume of oxygen would be required for the complete combustion of 10 cm^3 diphosphine?

c) Calculate the volume occupied by 0·66 g of diphosphine.

(Take the molar volume to be 24 litres mol^{-1})

Solution

a) This is just Knowledge and Understanding. 'Electronegativity' is the attraction of an atom for shared electrons in bonds. You simply have to learn this.

b) Remember, the volumes and gases are directly proportional to the number of moles. Two moles of diphosphine react with seven moles of oxygen. There are $3\frac{1}{2}$ times as many moles of oxygen as there are moles of diphosphine. Therefore, if you have 10 cm^3 of diphosphine, you need 35 cm^3 of oxygen.

c) The formula for diphosphine is P_2H_4. One mole weighs (2 × 31) + (4 × 1) = 66 g.

0·66 g is therefore 0·01 mole.

The volume is therefore 0·01 × 24 = **0·24 litre**.

Chapter 3

THE WORLD OF CARBON

Unit 2 is 'The World of Carbon'. There's no need to revise Standard Grade carbon chemistry, since Higher work replaces it.

The Unit comprises the following topics.

a) Fuels
b) Nomenclature and structural formulae
c) Reactions of carbon compounds
d) Uses of carbon compounds
e) Polymers
f) Natural products

Fuels

Summary

Petrol is a complex hydrocarbon mixture, made by reforming naphtha, a $C_5 - C_9$ fraction of crude oil. Naphtha is unsuitable as car fuel, but when reformed to a mixture of branched, cyclic and aromatic hydrocarbons, it burns well in a car engine.

The hydrocarbon blend in petrol is adjusted to take account of the seasonal temperature, ensuring that it is always volatile enough. For example, in winter, butane is added.

Premature ignition ('knocking') of the gas air mixture in the engine is eliminated by cyclic, aromatic and branched compounds.

Ethanol is an alternative fuel, usually added to petrol. It's made by fermenting materials like sugar cane and is therefore renewable, unlike petrol which comes from limited resources of fossil fuels.

Methanol is also an alternative. Like ethanol it's less volatile than petrol and so it's less dangerous in the event of fire. However, it can lead to corrosion of parts of the engine since methanol can absorb water.

Biogas is generated when bacteria degrade biological material in the absence of oxygen. Biogas is mainly methane. It is also a renewable fuel.

Hydrogen is a good fuel, and burns to produce water only, causing no pollution.

It can be made by electrolysis of water. Scientists are investigating the use of solar panels to generate the electricity required.

Figure 3.1 A hydrogen-powered car

Example 1

Which process produces aromatic hydrocarbons?

A Reforming of naphtha

B Catalytic cracking of propane

C Reforming of coal

D Catalytic cracking of heavy oil fractions

Solution

If you've taken in what you read above you should pick **A** as your answer.

Example 2

Carbohydrates like glucose can be used to generate biogas which is mainly methane.

State one advantage of biogas over natural gas.

Solution

Natural gas is a fossil fuel, and therefore it is non-renewable. Biogas can be made, not just from glucose, but from many other organic materials. It is therefore renewable.

Nomenclature and Structural Formulae

Summary

At Higher, you use numbers to show the positions of branches and multiple bonds.

If asked to name this molecule:

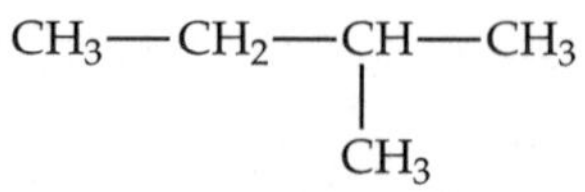

you can't say: 'Five carbons – pentane!'; you can't even say: 'A four carbon chain – butane!'. This molecule is **2-methyl butane**. Can you see why?

The same goes for alkenes – compounds with carbon to carbon double bonds.

$$CH_3—CH_2—CH=CH_2$$

The double bond is between carbon one and carbon two, so the correct answer is:

But-1-ene

The carbons are numbered from the end which gives the lowest numbers in the name. That's why both examples above are numbered from the right rather than the left.

If you forget how to name molecules using numbers, then turn to your trusty Data Booklet (page 6) where you'll find examples of alcohols and alkenes named with numbers. It probably won't answer the question you're tackling in your Higher paper, but at least it will remind you of how these compounds are named. It even shows you where to put the hyphens in the names, although you won't lose any marks for missing these out!

Example 3

The structure of a typical branched hydrocarbon found in petrol is shown.

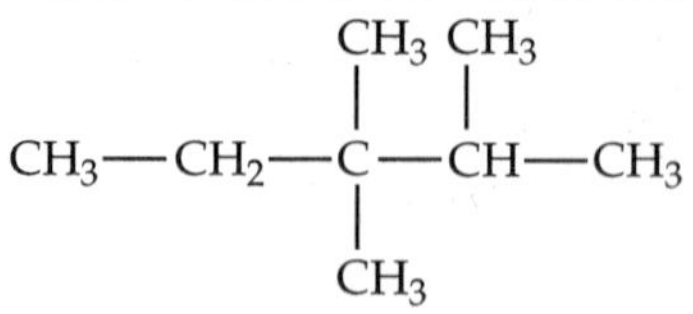

a) Name this hydrocarbon.

b) Name another type of petrol hydrocarbon which burns efficiently.

Solution

a) The longest continuous chain of carbon atoms in this molecule contains five atoms. This is therefore a pentane. It has three methyl groups ($-CH_3$) attached to it. It is therefore a trimethylpentane. You number it from the right because this will lead to lower numbers in the name. It is **2,3,3-trimethylpentane**.

b) Aromatic or cyclic

Functional groups

What you should know

You need to know these like the back of your hand! Learn them now! Here they are.

Group	Name	Found in
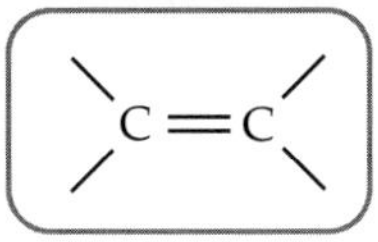	Double bond	Alkenes
$-C\equiv C-$	Triple bond	Alkynes
–OH	Hydroxyl group	Alcohols
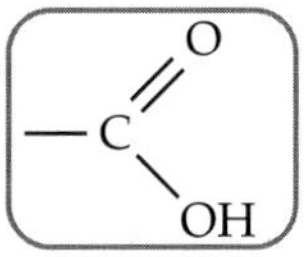	Carboxyl group	Carboxylic acids
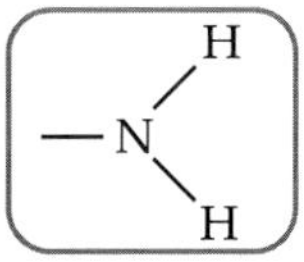	Amine or amino group	Amines
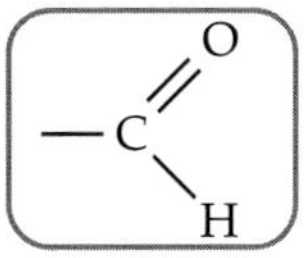	Carbonyl group	Aldehydes
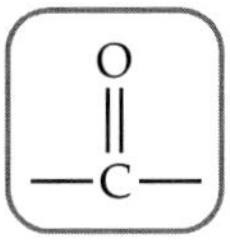	Carbonyl group	Ketones
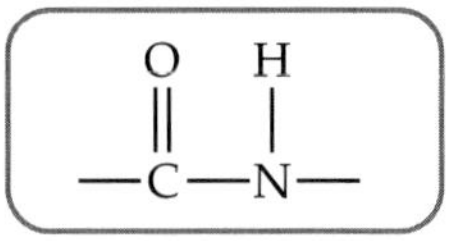	Amide/peptide link	Proteins/polyamides

What you should know *continued* ➢

What you should know continued

Group	Name	Found in
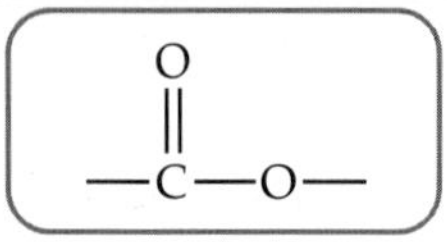	Ester link	Esters/polyesters
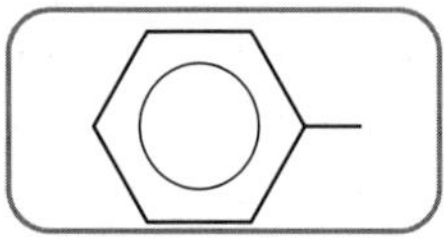	Phenyl group	Aromatics

Aromatic compounds have benzene rings and are very stable due to the presence of delocalised electrons.

Make sure you can read the benzene ring. There are six carbon atoms and six hydrogen atoms even if they're not shown! Anything attached to the benzene ring may mean fewer hydrogen atoms.

Example 4

The structural formula of the antiseptic, TCP, is shown in Figure 3.2.

a) What makes the benzene ring stable?

b) Write the molecular formula for TCP.

c) The systematic name for TCP is 2,4,6-trichlorophenol.

The systematic name for Dettol, another antiseptic, is 4-chloro-3,5-dimethylphenol.

Draw its structural formula.

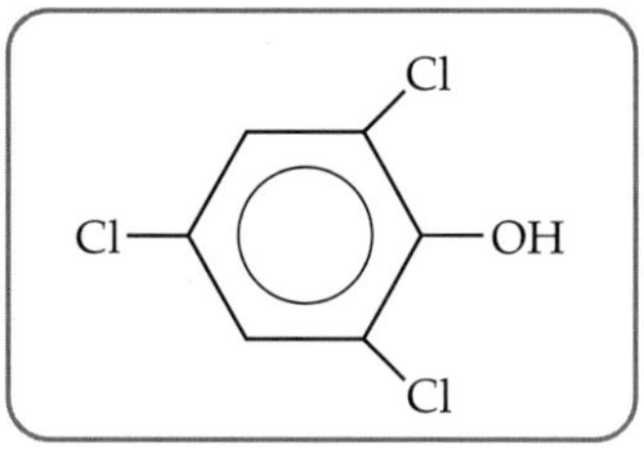

Figure 3.2 Structural formula of TCP

Solution

a) Easy question – delocalised electrons.

b) Benzene has six carbon and six hydrogens. This benzene ring has four groups attached to it. So it's lost four hydrogens. However, it's gained one because of the –OH group. So our answer is $C_6H_3Cl_3O$. It won't do to write $C_6H_2Cl_3OH$ because that's not a molecular formula. A molecular formula simply lists the types of atom and the numbers of the atoms. It doesn't tell you anything at all about the structure.

c) If the systematic name for TCP is 2,4,6-trichlorophenol, the carbon atom with the –OH must be carbon 1. So the structure you want for Dettol is as shown in Figure 3.3.

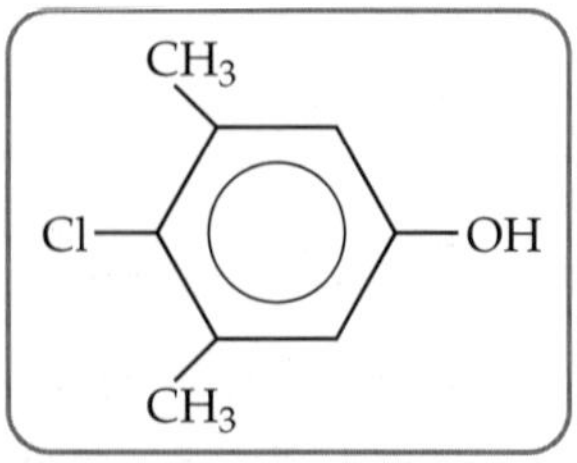

Figure 3.3 Structural formula of Dettol

Example 5

The structure of 4-methylphenol is:

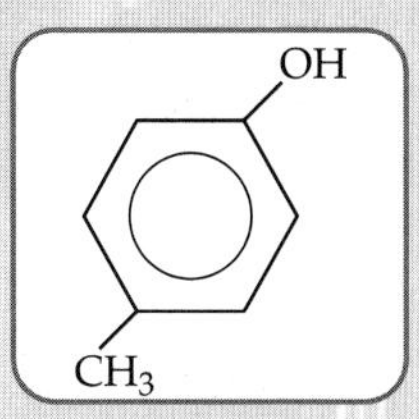

Write its molecular formula.

Solution

Again, you know that benzene has six carbon atoms. The methyl group provides a seventh. Benzene has six hydrogen atoms, but two have been replaced.

However, the $-$OH group adds one and the methyl group adds three hydrogens. The molecular formula is therefore C_7H_8O. An answer like $CH_3C_6H_4OH$ isn't a molecular formula – it's a shortened structural formula.

Drawing structural formulae

You have to be careful when you are drawing structural formulae. If you are asked to draw the **full** structural formula for a compound, then you have to show **all** the bonds. For example, suppose you're asked to draw the **full** structural formula for ethanoic acid, then this is what you have to draw.

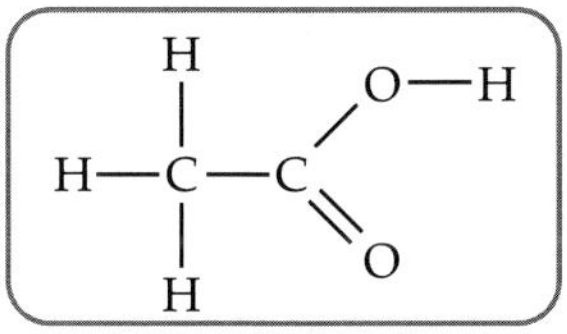

You might even get away with this

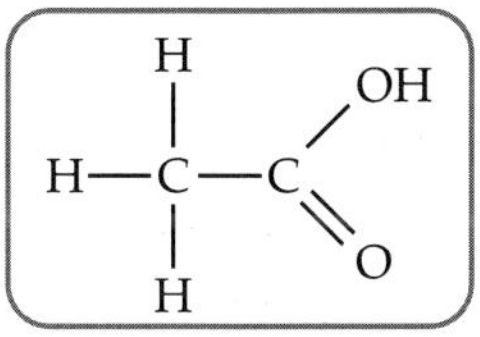

although it's not a **full** structural formula – and neither is this.

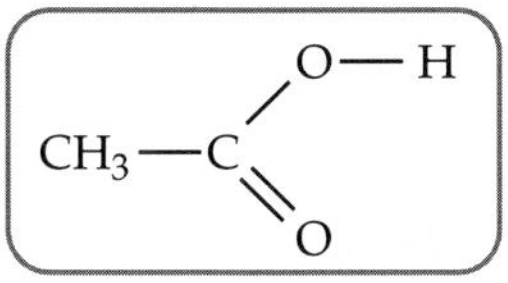

If a question says 'Draw **a** structural formula for.....' instead of 'Draw **the full** structural formula for...' then you can get away with grouping atoms together like $-$OH instead of $-$ O $-$ H or showing all the bonds in a methyl group, $-CH_3$. If your structural formula contains enough to convince the marker that you know the structure, you'll get the mark.

Reactions of Carbon Compounds

Type of reaction?

You'll find a couple of marks in the Higher paper for identifying reaction types.

What you should know

- **Condensation**: Molecules combine, releasing a small molecule (usually water) at each join. For example, when acids and alcohols form esters and when amino acids form proteins.
- **Hydrolysis**: The opposite of condensation. A larger molecule breaks apart as water is added to it. For example, when an ester breaks down to acid and alcohol or the breakdown of protein to amino acids.
- **Oxidation**: Addition of oxygen or removal of hydrogen, increasing the ratio of oxygen to hydrogen in a compound.
- **Reduction**: Removal of oxygen or addition of hydrogen, decreasing the ratio of oxygen to hydrogen in a compound.
- **Addition**: Double or triple bonds open and bond to other atoms. A double bond adds one mole of compound e.g. bromine. A triple bond adds two moles of compound. Addition of hydrogen is **hydrogenation**. (**Dehydrogenation** is the *removal* of hydrogen.)
- **Hydration**: This is addition of water to the double bond, as an H to one atom and an OH to the other. It can be used to convert an alkene to an alcohol.
- **Dehydration**: The opposite of hydration, often catalysed by aluminium oxide. Actual products depend on where the double bond is located in the molecule.
- **Addition polymerisation**: This type of polymerisation results from addition of molecules across double bonds. It can only occur when the monomer has a $C=C$ double bond.
- **Condensation polymerisation**: This type of polymerisation occurs when monomers have two functional groups, enabling condensation reactions to take place at each end of the molecule.
- **Cracking**: This involves long chain alkanes being broken down to shorter alkanes and alkenes. Heat and a catalyst (often aluminium oxide) are required.
- **Reforming**: This is really a set of reactions, which rearrange the atoms in an organic molecule, without greatly altering the number of carbon atoms.

Here are some examples for you to try!

Example 6

Which of the following equations shows a reaction involved in reforming?

A $C_6H_{14} \rightarrow C_6H_6 + 4H_2$

B $C_4H_8 + H_2 \rightarrow C_4H_{10}$

C $C_2H_5OH \rightarrow C_2H_4 + H_2O$

D $C_8H_{18} \rightarrow C_4H_{10} + C_4H_8$

Solution

There's a lot of chemistry tucked into this question. As usual, the best way of tackling it is to know that reforming rearranges the atoms in a molecule, without changing the number of carbon atoms. That only lets you pick **A** for the answer and this is the right choice. However, it's always as well to eliminate the other contenders! B is an addition reaction (or hydrogenation). C is dehydration, and D is cracking.

Example 7

What type of reaction occurs when propanol becomes propene?

A Condensation

B Dehydration

C Hydration

D Hydrolysis

Solution

If you know that propanol is $CH_3CH_2CH_2OH$ and propene is C_3H_6 then it should be easy to see that propanol has lost two H atoms and one O atom, so the answer must be **B** – dehydration.

Example 8

What kind of reaction is involved in the reaction below?

methanol → methanal

Solution

If you spot that methanol is an alcohol and methanal is an aldehyde, you should know that this is oxidation.

If you know that the formula for methanol is CH_3OH and the formula for methanal is HCHO, then you may see that the oxygen : hydrogen ratio has switched from 1 : 4 in methanol to 1 : 2 in methanal, which is the same as a change from 0·25 to 0·5 – an increase – so this reaction must be oxidation. Methanal is CH_2O but it's not seawater! Think about it!

Example 9

What is the following type of reaction?

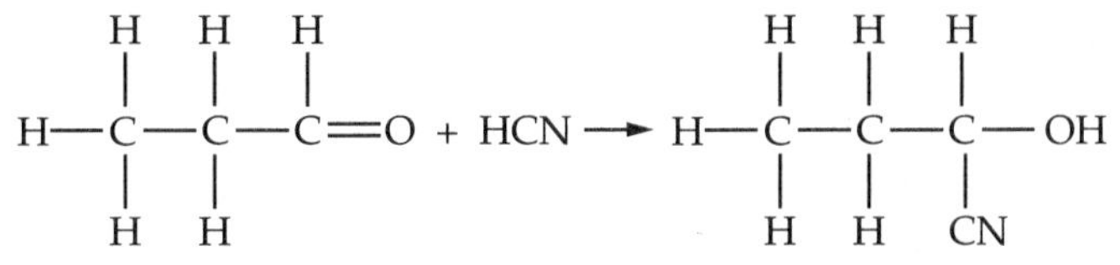

Solution

The propanal on the left-hand side has a double bond – the product on the right-hand side has no double bond. So this has to be an example of **addition**. It's Problem Solving, because you haven't met addition to bonds other than the carbon to carbon double or triple bond. You also haven't met HCN as a reactant in addition reactions.

Primary, secondary and tertiary alcohols

Chemists classify the position of a hydroxyl group in an alcohol molecule as primary, secondary or tertiary.

- In a primary alcohol, the –OH group is at the **end** of a chain.

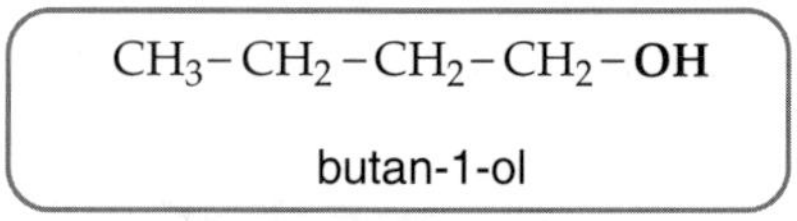

Figure 3.4 A primary alcohol

- In a secondary alcohol, the –OH is somewhere along the chain – not at the end. Like in Figure 3.5.

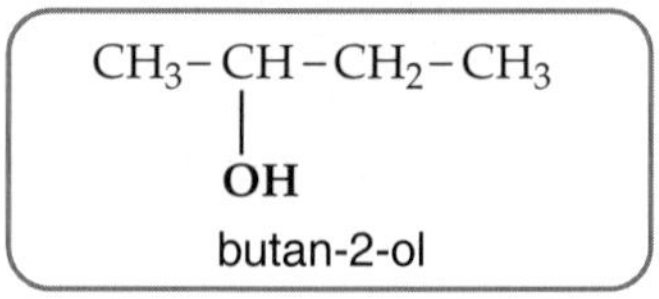

Figure 3.5 A secondary alcohol

- In a tertiary alcohol, the –OH is at a branch point on the chain. Like in Figure 3.6.

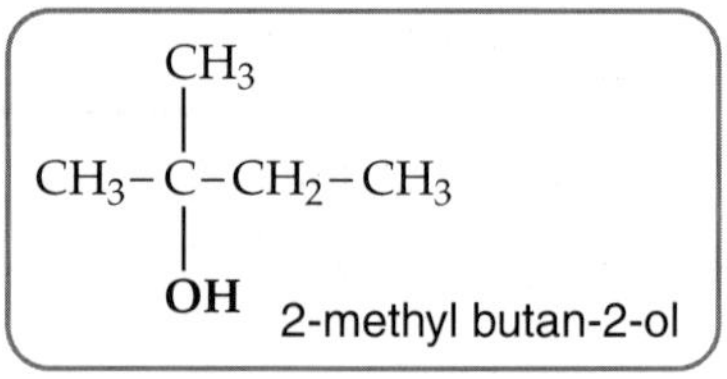

Figure 3.6 A tertiary alcohol

When you oxidise these alcohols, the primary ones become **aldehydes** and the secondary ones become **ketones**.

Aldehydes have a carbonyl group (=O) at the end of the molecule (where you find the oxygen in a primary alcohol). Ketones have a carbonyl group in the middle of the molecule (where you find the oxygen in a secondary alcohol). The oxidation of a primary alcohol results in the aldehyde formed being oxidised further to a carboxylic acid. So you have:

$$\text{primary alcohol} \underset{\text{reduce}}{\overset{\text{oxidise}}{\rightleftharpoons}} \text{aldehyde} \underset{\text{reduce}}{\overset{\text{oxidise}}{\rightleftharpoons}} \text{carboxylic acid}$$

$$\text{secondary alcohol} \underset{\text{reduce}}{\overset{\text{oxidise}}{\rightleftharpoons}} \text{ketone}$$

The oxidation of a primary alcohol to an aldehyde involves this change.

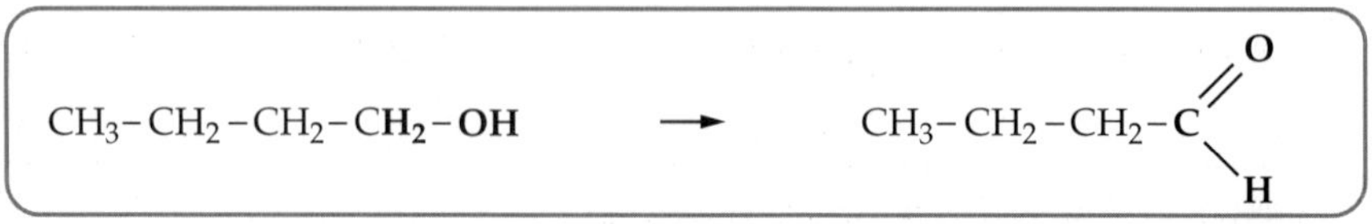

Example 10

The oxidation of 4-methylpentan-2-ol to a ketone results in a mole of alcohol:

A losing 2 g

B gaining 2 g

C gaining 16 g

D losing 16 g.

Solution

If you know that turning an alcohol into an aldehyde or a ketone involves removing two hydrogen atoms from the alcohol, then it should be clear that the answer is **A**. You might be tempted to say that this is oxidation, and pick C, because gaining 16 g implies adding oxygen. It's essential to learn what happens in these oxidation reactions!

Example 11

Here are some steps in a complex reaction.

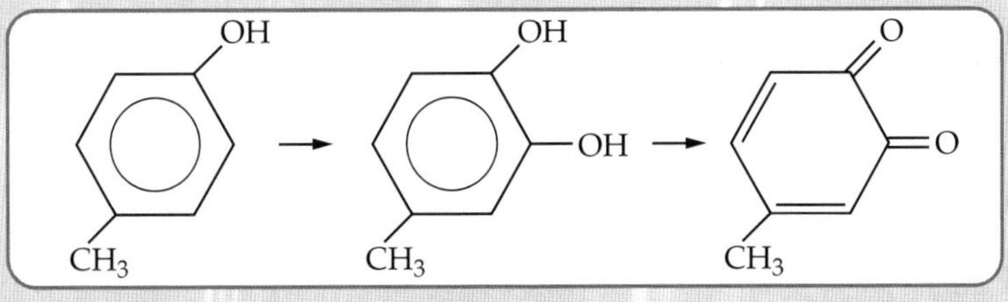

The same type of reaction takes place in **both** steps.

What type of reaction is involved?

Solution

This should be quite easy to deal with. You can see that the hydroxyl (—OH) groups in the middle molecule have ended up as carbonyl (C=O) groups in the final molecule. That is, an alcohol has become a ketone – this is an example of **oxidation**.

Does this mean that the first reaction is oxidation too? Let's work out the molecular formulae of the first and second molecules.

The first is C_7H_8O. Oxygen to hydrogen ratio is 1 : 8

The second is $C_7H_8O_2$. Oxygen to hydrogen ratio is 2 : 8

There has been an increase in the oxygen : hydrogen ratio, so the reaction is **oxidation**.

Example 12

Some reactions of a triglyceride are shown below.

$CH_2OOCC_{17}H_{35}$		$CH_2OOCC_{17}H_{35}$		CH_2OH		$C_{17}H_{35}COOH$
$CHOOCC_{17}H_{31}$	$\xrightarrow{P}$	$CHOOCC_{17}H_{35}$	$\xrightarrow{Q}$	$CHOH$	+	$C_{17}H_{35}COOH$
$CH_2OOCC_{13}H_{25}$		$CH_2OOCC_{13}H_{27}$		CH_2OH		$C_{13}H_{27}COOH$

Identify the types of reaction taking place at P and Q.

Solution

You should be able to spot that in reaction **P** the product contains more hydrogen atoms than the reactant. This must therefore be an example of **hydrogenation**. Could it also be an example of addition? This would mean that the reactant would have to have carbon to carbon double bonds. The $-C_{17}H_{31}$ group must have **two** double bonds – 17 carbon atoms need 35 hydrogens to make a saturated molecule, and there are only 31. The $-C_{13}H_{25}$ group must have **one** double bond – 13 carbon atoms need 27 hydrogens to make a saturated molecule. The reaction must involve hydrogen adding to the double bonds, so this is also an example of **addition**.

How about reaction Q? You should be able to recognise the structure of the alcohol **glycerol** as a product of this reaction, and there are three molecules with **carboxyl** groups – this means there are three **acid** molecules.

The reactant of reaction Q contains the pattern of atoms $-OOC-$ in its structure. You should recognise this as an ester. So you have an ester becoming an alcohol and some acid molecules. The ester molecule has been broken open, and hydrogen and oxygen atoms have been added to the resulting bits. This is therefore an example of **hydrolysis**.

Making it happen – reagents and catalysts

What you should know

- **Reagents** are chemicals which cause particular chemical changes. Make sure you know them and what they do.
- **Catalysts** also cause chemical changes, but aren't consumed in the process.
- **Esterification (a condensation reaction)**: The catalyst is **concentrated** sulphuric acid. Miss 'concentrated' and miss out on half a mark
- **Hydrolysis**: These reactions are catalysed by acids, alkalis or enzymes.
- **Oxidation**: Learn the two important oxidising agents – **acidified** potassium dichromate and **hot** copper oxide. If you miss out 'acidified' or 'hot' you'll miss out on half a mark.
- **Addition to $C=C$**: Bromine, chlorine, iodine, hydrogen bromide, chloride or iodide could be involved. Hydrogen (in hydrogenation) or water (in hydration) can also be involved.
- **Dehydration**: Aluminium oxide catalyses removal of water from an alcohol, forming an alkene. The products depend on where the $-OH$ group is located in the molecule.
- **Cracking**: This requires heat and the catalyst aluminium oxide.

Example 13

Some reactions of carbon compounds are shown in Figure 3.7.

a) i) Name compound Q.

ii) What term is used to describe compounds such as hot copper oxide and acidified potassium dichromate?

b) Name reaction A.

c) The product of reaction B is an ester. Name this ester.

d) If the alcohol used had been propan-2-ol, how would the product of reaction A have been affected?

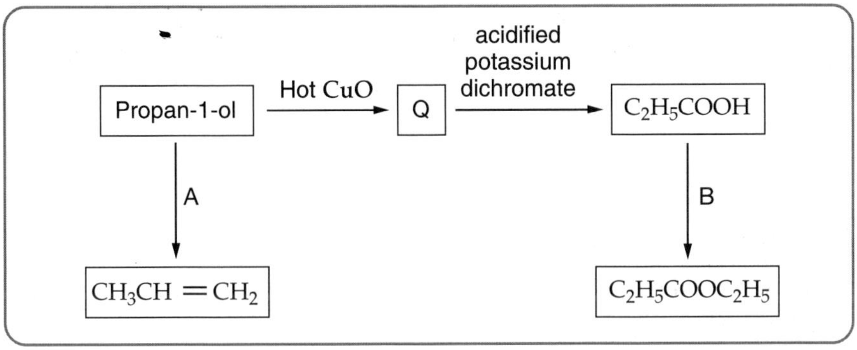

Figure 3.7 Reactions of carbon compounds

Solution

a) i) Since this set of reactions begins with an alcohol (propan-1-ol) and ends with a carboxylic acid (C_2H_5COOH), the only possible intermediate is an aldehyde, in this case propanal.

ii) These are oxidising agents. You should know their names.

b) Reaction A has converted an alcohol to an alkene. This is dehydration. If in any doubt about this, you could write the formula for propan-1-ol, C_3H_7OH, and note that the product has two hydrogen atoms and one oxygen atom fewer than the reactant.

c) Ethyl propanoate

d) It makes no difference in this case whether the alcohol is propan-1-ol or propan-2-ol. There is only one possible location for the double bond which results.

Example 14

Compound X is a secondary alcohol with the following structure.

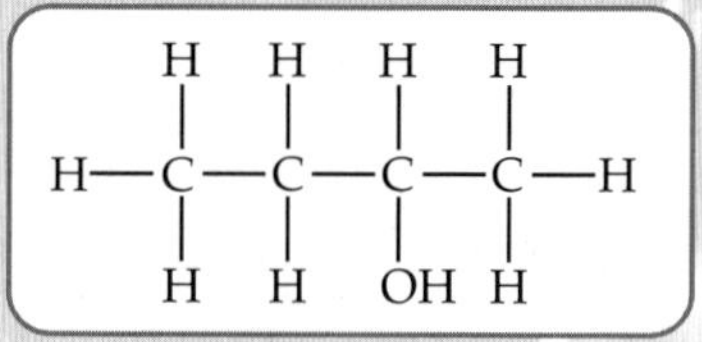

a) Name compound X.

b) Draw a structural formula for the tertiary alcohol which is an isomer of compound X.

c) When passed over hot aluminium oxide, compound X is dehydrated, producing isomeric compounds Y and Z.

Both Y and Z react with hydrogen bromide, HBr. Y gives two products while Z gives only one.

Name compound Z.

Solution

a) Butan-2-ol. If you're not sure where '2' goes, you'll find the Data Booklet gives you examples of the names of alcohols with numbers.

b) In a tertiary alcohol, the —OH group is attached to a carbon atom which in turn is attached to three other carbon atoms.

This is the structure you want.

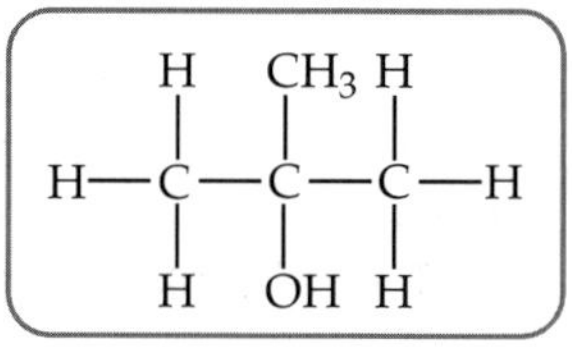

It's an isomer of compound X – it's got four carbon atoms, ten hydrogen atoms and one oxygen atom, so it's an isomer of butan-2-ol.

The structure drawn above is acceptable because the question didn't say 'Draw the **full** structural formula ...' What makes this question quite difficult is the language – you need to know meanings of the words **'tertiary'** and **'isomer'**.

c) This is really Problem Solving. Dehydration removes water, in the form of the OH and a neighbouring H. The H can come from the left or the right of OH, resulting in **two** possible products both of which are alkenes.

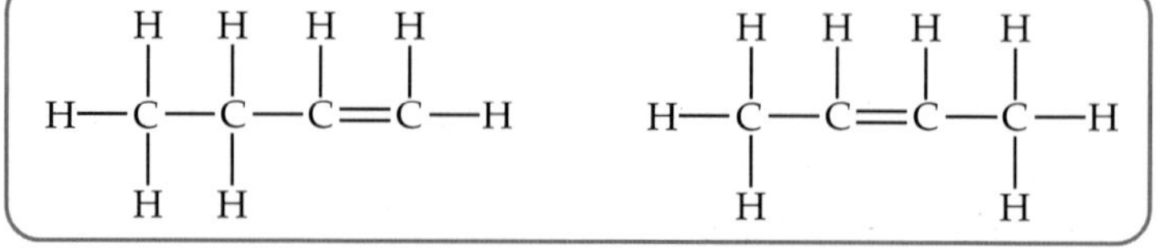

Which is Y and which is Z?

Z reacts with HBr to produce just one product. H will add to one end of the double bond and Br will add to the other. The left-hand molecule will form two possible products, 1-bromobutane and 2-bromobutane. The right-hand molecule will give just one product. It doesn't matter which end of the double bond the bromine adds to, the product is always 2-bromobutane.

So Z is the right-hand molecule, **but-2-ene**.

Example 15

A compound has the following structural formula.

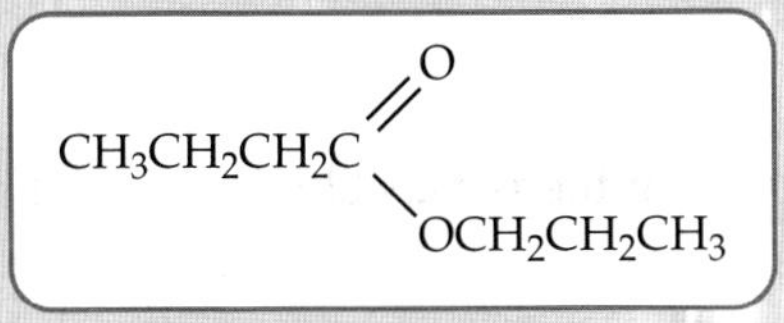

It can be made from:

A propanol and propanoic acid
B propanol and butanoic acid
C butanol and propanoic acid
D butanol and butanoic acid.

Solution

In this kind of question, it's a good idea to count carbon atoms in the structure. There are seven atoms. In which case your answer will also have to contain seven carbon atoms.

A contains six carbons, B and C both contain seven, and D contains eight. B and C are possible answers. The carbon atom with the doubly bonded oxygen belongs to the acid. That means the acid has four carbons (butanoic acid) so the answer is **B**.

It's handy to realise that different esters and simple carboxylic acids having the same number of carbon atoms are all isomers of each other. So hexanoic acid, methyl pentanoate and propyl propanoate are all isomeric. They all contain six carbon atoms.

Example 16

An ester has the following structural formula.

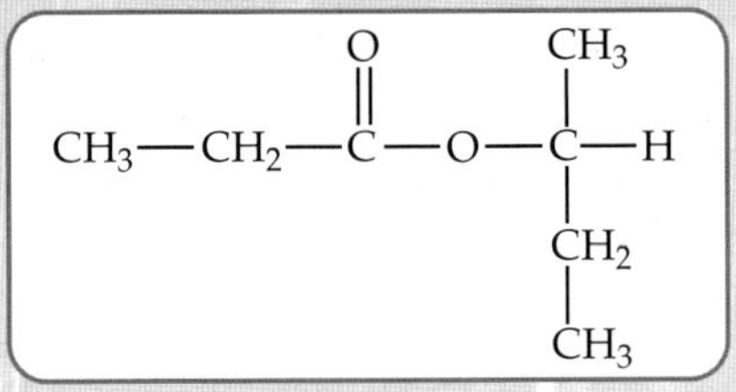

Hydrolysis of this ester would produce:

A propanoic acid and propan-1-ol
B propanoic acid and propan-2-ol
C butanoic acid and propan-2-ol
D propanoic acid and butan-2-ol.

Solution
Again, you can check this out by adding up the carbon atoms.

There's a total of seven carbon atoms. The molecules in A and B only amount to six, in each case. You can ignore these and look at C and D. If you read the left-hand side of the molecule as the acid part (note the C=O) then the three carbon atoms tell you that propanoic acid is involved. A quick check of the right-hand side should convince you that butan-2-ol is also involved. So the correct answer is **D**.

Uses of Carbon Compounds

Summary

You haven't much to learn. You are more likely to be asked about the uses of esters than of any other type of compound. The uses of esters are as *flavourings*, *perfumes* and *solvents*. Don't give any other use for a simple ester.

Polymeric esters are used in textile fabrics and resins.

Figure 3.8 Lots of esters here

Example 17

Which statement can be applied to polymeric esters?

A They can be used as flavourings, solvents and perfumes.

B They are condensation polymers made from amino acids.

C They are used as textile fabrics and resins.

D They are cross-linked addition polymers.

Solution
Not A – that's simple esters.

Not B – amino acids are used to form proteins.

C – you should **know** this!

Not D – esters are formed by a condensation process, not addition.

Benzene and related compounds are used as **feedstocks** in the chemical industry – that is, they are used to make other compounds of carbon such as medicines, dyes and herbicides.

Halogenoalkanes (alkanes with hydrogen atoms replaced by halogen atoms) include CFC's, once popular as aerosol propellants and also once used in fridges. A lot of them are useful solvents and were once used in products like Tippex.

Figure 3.9 Nobody wants them now

Polymers

Addition polymers

You met addition polymerisation in Standard Grade and learned how to draw a section of poly(ethene) made from three ethene molecules.

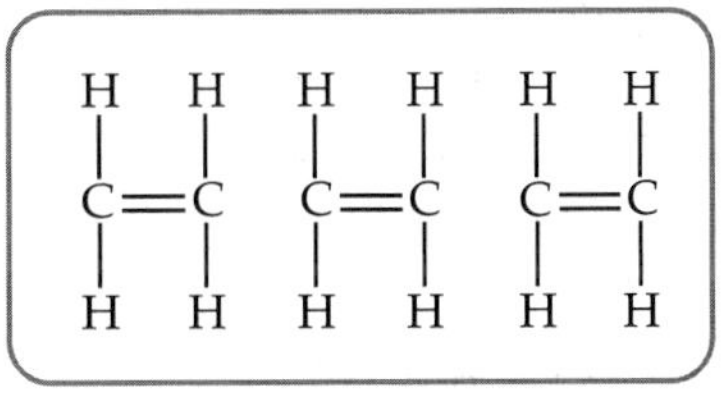

Figure 3.10 Three ethene molecules

The polythene section was like this (don't forget the open ends).

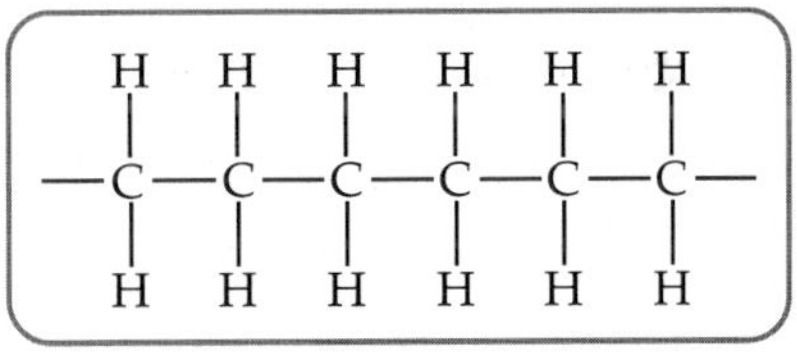

Figure 3.11 Section of polyethene

At Higher, it's a bit more challenging. For example:

Question

Vinyl acetate has this structural formula.

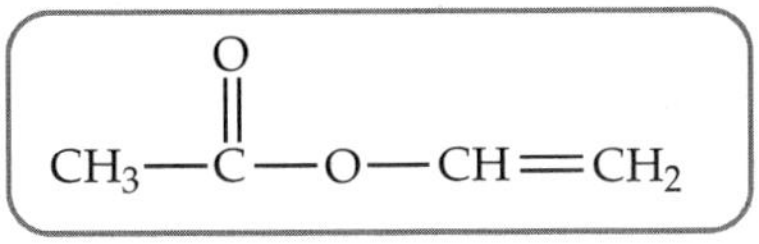

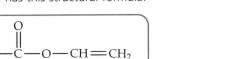

Figure 3.12 Vinyl acetate molecule

Draw a section of the poly(vinyl-acetate) molecule showing three monomer units.

To do this, you have to rearrange the molecule to form an 'H' shape using the double bond as the cross bar. Like this:

It may not look like the molecule in the question, but it's identical!

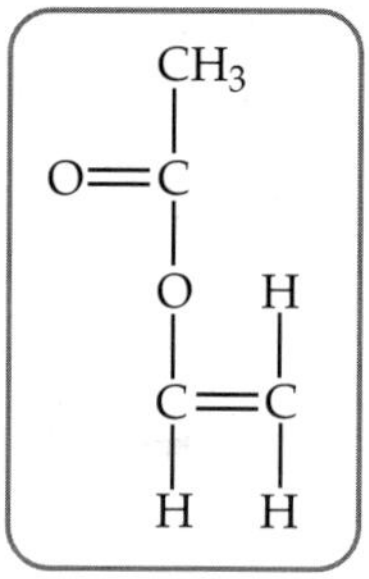

Figure 3.13 Rearrange vinyl acetate molecule

So the answer looks like this:

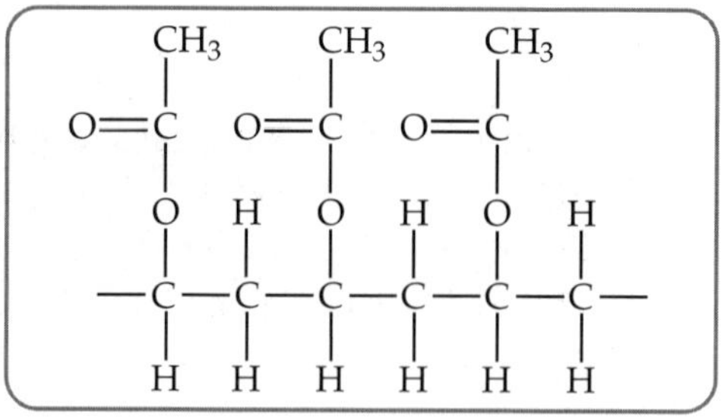

Figure 3.14 Section of poly(vinyl acetate)

Example 18

Part of a polymer molecule is shown below.

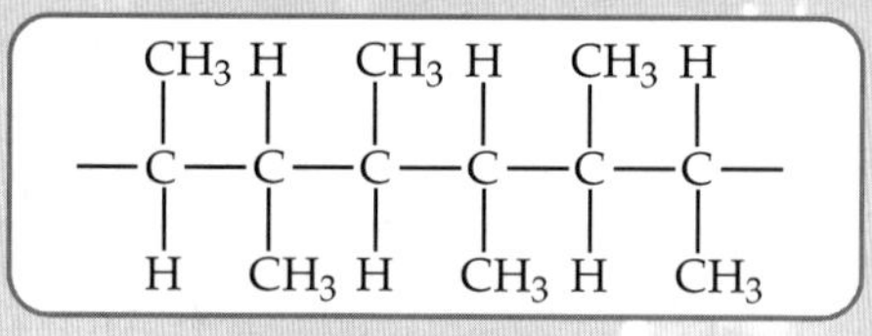

Name the monomer used.

Solution

If you split the backbone into two-carbon units you have:

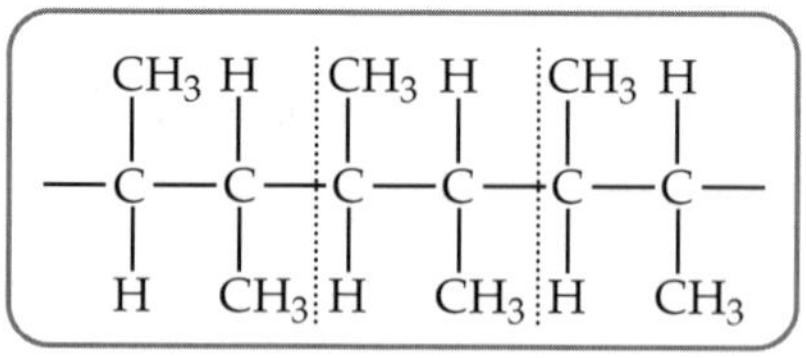

Then it should be obvious that the monomer has four carbon atoms with a double bond in the middle. That is, the answer is but-2-ene, $CH_3-CH=CH-CH_3$.

Condensation polymers

These form when monomers have **two** functional groups, enabling a reaction at **each end** of the molecule.

They're easy to spot, because their 'backbone' is not made of carbon atoms only, as with addition polymers. Many have ester links (as in polyesters) or amide links (as in polyamides like nylon, and Kevlar). (See page 31 for these links.)

Polyesters and polyamides are used to make fibres, for clothing. Polyesters can also be made with 'cross-links' between polymer chains. This makes them into rigid resins.

New Polymers

What you should know

You don't need to know a lot about these but you should know one or two facts about each.

- **Kevlar** is like nylon, except that is aromatic – it's based on benzene rings – and has loads of hydrogen bonding which gives it great strength in relation to its weight. It's used to make car tyres, body armour and to reinforce aircraft fuel tanks.
- **Polyethenol** is soluble in water. It's got —OH groups on it so it forms hydrogen bonds with water, making it soluble. It's used to make surgical stitches inside the body. In time, they dissolve. It's made by a process called 'ester exchange'.

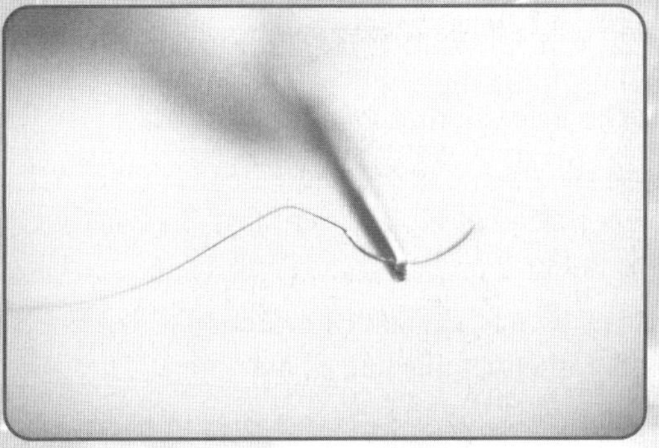

Figure 3.15 Soluble stitches don't have to be removed later

- **Poly(vinyl carbazole)** conducts electricity if light strikes it. It's used in photocopiers.
- **Poly(ethyne)** can be processed into an electrical conductor.

  Practise showing how it forms from ethyne.

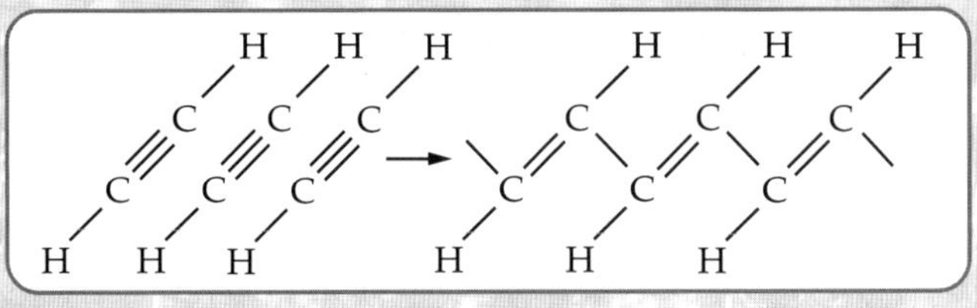

Figure 3.16 Formation of poly(ethyne) from ethyne

- **Biopol** is biodegradable, and polythene can be modified to become biodegradable by putting C=O groups into its carbon backbone.

Natural Products

The natural products dealt with are fats, oils and proteins.

Fats and Oils

Three main types of fats and oils are: animal (e.g. beef fat), vegetable (e.g. sunflower oil), and marine (e.g. cod liver oil). Fats are solid at room temperature and oils are liquids. They provide the body with a concentrated energy source.

Fats and oils are esters. The alcohol involved is glycerol, $CH_2OHCHOHCH_2OH$. This is a colourless liquid, thick because of lots of hydrogen bonding at the $-OH$ groups. The carboxylic acids involved have even numbers of carbon atoms from C_4 to C_{24}, but usually C_{16} and C_{18}. Learn the formulae for stearic acid and oleic acid.

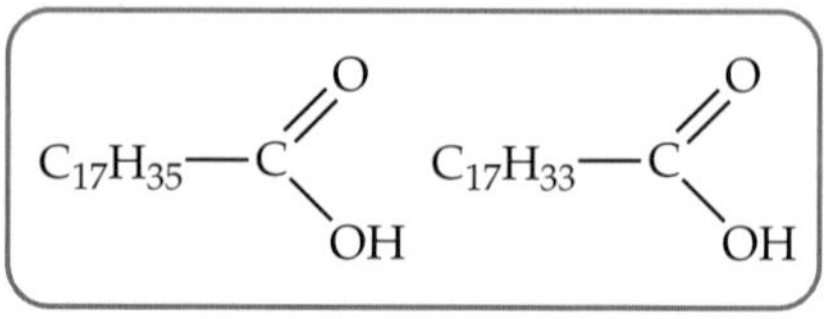

Figure 3.17 Formulae for stearic acid and oleic acid

These are often called 'fatty acids'.

Stearic acid is saturated. Oleic acid is unsaturated, with a carbon to carbon double bond. Fats form when one mole of glycerol forms an ester (condenses) with three moles of saturated fatty acids. Oils involve unsaturated fatty acids.

Oils are liquids because the double bonds cause bends in the molecule so the molecules can't get as close to each other as molecules of fats. The van der Waals forces in oils are therefore weaker than in fats.

Oils are converted to fats by the addition of hydrogen to the double bonds, a process known as hardening and used in the manufacture of margarine.

Hydrolysis of fats produces three moles of fatty acids to one of glycerol. If the hydrolysis involves excess alkali (e.g. NaOH) the alkali reacts with the stearic acid, giving **sodium stearate** which is **soap**!

Example 19

Which of the following decolourises bromine solution least quickly?

A Sunflower oil

B Oct-1-ene

C Cod liver oil

D Beef fat

Solution

You should have no problem with this one! Just remember that oils are made from unsaturated fatty acids and fats are made from saturated fatty acids. The fats contain no carbon to carbon double bonds and will not react quickly with bromine. **D** is the answer.

Proteins

Proteins make up our skin, hair, muscle and all the enzymes in our bodies. They form important foods including meat, fish, eggs, peas and beans. They comprise large molecules built of smaller molecules called

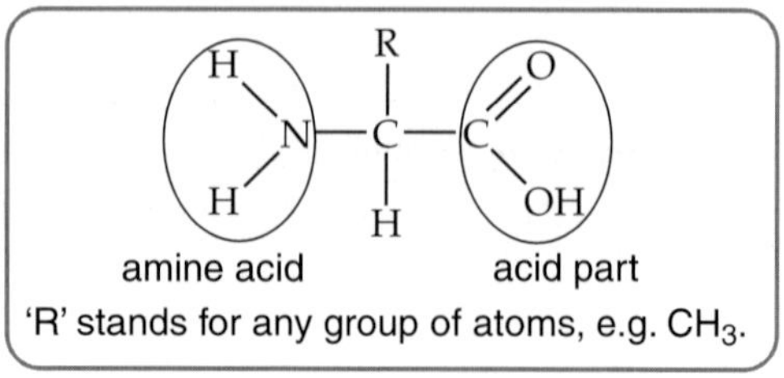

Figure 3.18 Structure of an amino acid

amino acids. The structure of an amino acid has an amine part and an acid part.

Proteins are condensation polymers of amino acids; as the amino acid units join up, water molecules are set free resulting in the formation of peptide links.

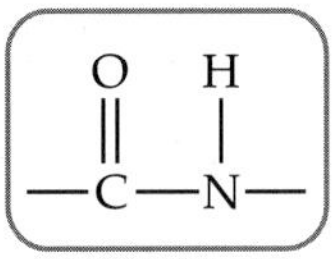

Figure 3.19
Peptide link

Our bodies hydrolyse food proteins into amino acids, and recombine them to form body proteins. Our bodies can make many amino acids. But there are some our bodies can't make. These **essential amino acids** have to be included in our diet.

Protein molecules are flexible. They can fold into many shapes, held in place by hydrogen bonds between the polar C=O and N−H groups along the chain.

Depending on how they're folded, they can form **fibrous** or **globular** proteins.

- Fibrous proteins form long, thin strong bundles. They're involved in structures of the body like skin, hair and muscles.
- Globular proteins are rounder and are involved in the maintenance and regulation of life processes. They include haemoglobin responsible for carrying oxygen round the body, hormones, like insulin, and enzymes.

Example 20

Amino acids react to form protein molecules by:

A condensation

B hydration

C hydrogenation

D hydrolysis.

Solution

Make sure you read this question carefully. Any biological process involving amino acids, proteins, and fats and oils will almost certainly be either hydrolysis or condensation. So you could pick either A or D. The correct answer, of course, is **A**. It's easy if you've learned it. But make sure you don't take the question to mean 'Amino acids form from proteins ...'. Then you'd give D as the answer.

Example 21

Oils melt more easily than fats because:

A oils have fewer hydrogen bonds

B there are fewer cross-links in oils

C oil molecules are less saturated

D there are weaker van der Waals forces in fats.

Solution

The basic difference between oils and fats is the type of fatty acid involved. In oils, they are unsaturated, in fats, they're saturated. The C=C leads to bends in the unsaturated acid which in turn leads to poorer packing and weaker van der Waals forces. **C** is the answer.

Example 22

Our bodies contain an important fibrous protein called collagen.

a) Describe a difference between a fibrous and a globular protein.

b) Name the four elements present in all proteins.

Solution

a) Fibrous proteins are long and thin. Globular proteins are round (spherical).

b) Just remember – proteins are made up of amino acids. 'Amino' tells you that nitrogen and hydrogen have to be present. 'Acid' tells you that carbon and oxygen (and hydrogen) have to be present (from the carboxyl group). Or just remember 'CHON'.

Example 23

When amino acids condense, water molecules are set free, forming an amide link.

Which of the following shows this process?

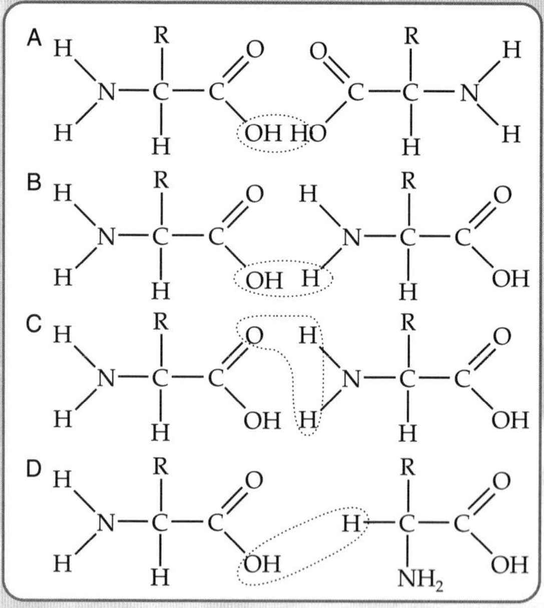

Solution

If you know what an amide link looks like, there's only one possible answer to this question.

This is an amide link:

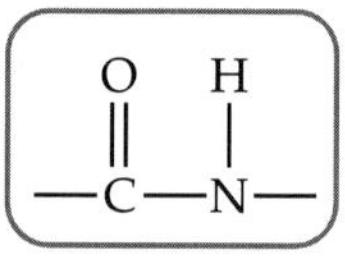

and you should see that only **B** can lead to such a structure.

Example 24

Olive oil has a number of uses.

a) Olive oil can be hardened to produce a fat. The catalyst used is nickel.
What type of catalyst is nickel in this reaction?

b) In what way does the structure of a fat molecule differ from that of an oil molecule?

c) Olive oil can be hydrolysed by sodium hydroxide to form sodium salts of fatty acids.
Name the other product of the reaction

d) Give a commercial use for sodium salts of fatty acids.

Solution

a) The catalyst is a solid, but the olive oil is a liquid. They are in different states, so it is a heterogeneous catalyst.

b) Fats are saturated and oils are unsaturated.

c) All oils and fats are based on glycerol and various fatty acids. Glycerol is always a product of the hydrolysis of oils and fats.

d) Soap!

Chapter 4

CHEMICAL REACTIONS

Unit 3 is called 'Chemical Reactions' but there aren't too many chemical reactions in it.

The Unit comprises the following topics.

a) The chemical industry
b) Hess's law
c) Equilibrium
d) Acids and bases
e) Redox reactions
f) Nuclear chemistry

The Chemical Industry

Summary

The chemical industry is one of the UK's largest, both in number of employees, and contribution to the economy.

The industry is described as 'capital intensive' rather than 'labour intensive'. This means that it employs fewer people than other industries which spend the same in building their factories.

A new industrial process starts with Research and Development, where background theory and the chemistry of the process are examined. Laboratory and computer modelling are involved. A pilot plant (a small scale version of the final process) lets the chemists and engineers see how the equipment will behave under operating conditions. Finally the bulk process is set up using a full scale plant.

Raw materials

The chemical industry uses raw materials – substances taken unchanged from the Earth's resources such as air, sea water, coal, crude oil, natural gas, and metal ores.

Feedstocks

Raw materials are turned into feedstocks – the actual reactants. The Haber process is a good example of this. The equation for the reaction is:

$$N_2 + 3H_2 \rightarrow 2NH_3$$

The nitrogen and hydrogen are feedstocks. However, the raw materials are mainly air (for the nitrogen) and natural gas (for the hydrogen).

Figure 4.1 Compare the size of this pilot plant with BP Grangemouth

Batch and continuous processes

A chemical plant can operate on a batch system or continuously. In a batch system a set of feedstocks is put in and converted to product. The process is repeated as often as required. In a continuous system, the feedstocks are constantly put in and product is constantly drawn off.

A batch operation allows the same plant to make different products and it's cheaper to build. A continuous process needs a smaller workforce and is cheaper to operate.

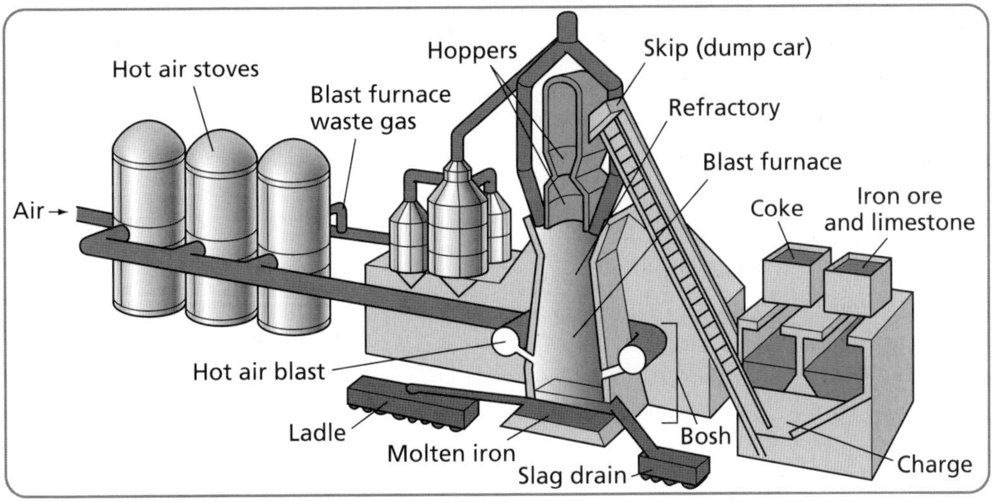

Figure 4.2 24/7 operation

Capital, fixed and variable costs

Capital costs refer mainly to the cash invested in building the plant.

Fixed costs must be met, whether or not the plant is operating. This includes repayment of the loan taken out to build the plant, council tax and salaries.

Variable costs include the cost of raw materials – if the plant is operating at full capacity, it needs more raw materials than if it's operating below full capacity. Distribution costs are variable too. It depends how much product has to be distributed.

Safety and the environment

Safety is a major concern in the industry. A lot of processes involve high temperatures and pressures, and toxic or flammable materials so care is taken to reduce the risk of accidents. It's important to take the quality of the environment into account when building and operating a chemical plant. Companies now find ways of using waste, rather than just dumping it or releasing it into rivers or the atmosphere. In the steel industry, slag was once a waste product but now it's made into building materials.

Example 1

Which of the following is **not** a raw material?

A Air

B Natural gas

C Ethene

D Water

Solution

Remember, raw materials are substances that you can find from the Earth's resources without having to change them in any way. Air, natural gas and water are all raw materials. Ethene **isn't** found naturally – it has to be made by cracking crude oil (which **is** a raw material). So **C** is the answer.

Example 2

Which of the following costs is a variable cost?

A Land rental

B Plant construction

C Labour

D Raw materials

Solution

Fixed costs are costs which can't be easily changed. Land rental is paid to the owner of the land on which the plant is built. This is probably paid at a fixed annual rate. The cost of plant construction is also fixed. Once a loan has been taken out to build the plant, the repayment can't be easily changed. In a similar way, labour costs are fixed. They can't be changed unless the actual number of employees is altered. You're left with the cost of **raw materials**. This will alter, depending on the demand for the plant's product. The more product required, the more raw materials required.

Example 3

Chlorine can be manufactured by different processes.

In the Castner Kellner method, sodium chloride solution (brine made from underground salt deposits) is electrolysed with graphite positive electrodes and a flowing mercury negative electrode. Sodium forms and dissolves in the mercury electrode, while chlorine forms at the graphite electrode and is piped off.

In the Deacon process, the following reaction is used.

$$4HCl + O_2 \rightarrow 2Cl_2 + 2H_2O$$

Example 3 *continued* ➢

Example 3 continued

a) Why is graphite able to conduct electricity?

b) In the first process, the solution of sodium in mercury is treated with water to give two valuable products.

Name the two products.

c) Suggest why the Deacon process is less economical than the Castner Kellner process.

Note: this question doesn't just draw from one unit – it has Unit 1, Standard Grade and Unit 3 material in it.

Solution

a) Graphite has delocalised electrons – electrons which can move freely throughout the graphite, just as in a metal.

b) This is Standard Grade work. It's not covered in Higher at all.

When a metal, high in the electrochemical series, like sodium, reacts with water, the products are hydrogen and the metal hydroxide, in this case, sodium hydroxide.

c) You could suggest a number of good reasons. The Castner Keller process uses a single reactant, brine, which is a raw material. Brine is also inexpensive. The Deacon process uses two reactants, hydrogen chloride and oxygen from the air. While air is cheap enough, hydrogen chloride has to be made and is quite expensive.

In the Castner Kellner method, the chlorine forms at the positive electrode and is led away. In the Deacon process, chlorine and water vapour form as a mixture of products and would have to be separated.

Example 4

Which of the following is produced by a batch process?

A Sulphuric acid from sulphur and oxygen

B Aspirin from salicylic acid

C Iron from iron ore

D Ammonia from nitrogen and hydrogen

Solution

You learned about ammonia and iron in Standard Grade. In the Haber process, nitrogen and hydrogen are fed steadily over an iron catalyst. Unreacted nitrogen and hydrogen are recycled into the system. You also learned about the blast furnace, which has iron ore and coke fed in at the top while iron and slag are removed at the bottom. These are both continuous processes. It's not clear how you choose between A and B. In fact, sulphuric acid is also made by a continuous process in which sulphur and oxygen are fed steadily into the plant.

Aspirin is made by a batch process. You can work this out by assuming that far smaller amounts of aspirin are made than of sulphuric acid, a major industrial chemical.

Hess's law

We'll deal with this in the 'Calculations' chapter!

Equilibrium

Summary

The basic idea here is that virtually all reactions are reversible. By reversible we mean that just as reactants can become products, products can become reactants.

In the Haber process for ammonia, the temperature mustn't be too high, otherwise the ammonia breaks down to give the original hydrogen and nitrogen.

We show it like this.

$$N_2 + 3H_2 \rightleftharpoons 2NH_3$$

At the start of the process, there isn't much product, so the reverse reaction is slow. As product accumulates and reactants are consumed, the forward reaction slows down and the reverse reaction speeds up. Eventually, the rate of forward and reverse reactions become equal. The reaction has reached 'equilibrium'. At equilibrium, the concentrations of reactants and products remain steady though not usually the same as each other. Sometimes it's called a 'dynamic equilibrium' because although the concentrations remain steady, there is constant interconversion of reactants and products.

Factors affecting equilibrium position

To encourage a ΔH negative reaction, use a low temperature.

To encourage a ΔH positive reaction, use a high temperature.

You can encourage the forward reaction by increasing reactant concentration, or by removing products as they form (preventing the reverse reaction).

If a reaction involves gases, a pressure increase moves the equilibrium to the side with smaller volume – that is, fewer moles of gas. A decrease in pressure has the opposite effect.

Catalysts don't affect the position of equilibrium although equilibrium is reached faster.

Figure 4.3 Performing seals know about equilibrium!

Example 5

Which statement about catalysts is **incorrect**?

A The position of equilibrium is unchanged.

B The position of equilibrium shifts to the product side.

C The rate of the forward reaction increases.

D The rate of the reverse reaction increases.

Solution

This should be a really easy question! You've picked **B**! That's right – a catalyst has no effect on the position of equilibrium.

Example 6

Iodine monochloride and chlorine gas take part in the following equilibrium.

$$ICl(l) + Cl_2(g) \rightleftharpoons ICl_3(s)$$

ΔH for the forward reaction is $-106\,kJ\,mol^{-1}$.

Which line in the table correctly identifies the conditions which will cause the greatest increase in the amount of solid in this equilibrium?

	Temperature	Pressure
A	low	low
B	low	high
C	high	low
D	high	high

Solution

This reaction is exothermic. It tries to get rid of heat. This is encouraged by a **low** temperature, so you pick either A or B.

There is one mole of gas on the left of the equation. The product is solid, so it has a smaller volume than the reactants. A smaller volume is achieved using a **high** pressure, so you are left with **B** as the answer.

Example 7

When chlorine is added to water, the following equilibrium is set up.

$$Cl_2(g) + H_2O(l) \rightleftharpoons 2H^+(aq) + ClO^-(aq) + Cl^-(aq)$$

What is the effect on the position of equilibrium of:

a) increasing the pressure

b) adding sodium chloride

c) adding sodium hydroxide?

Solution

In (a) you should notice that there is 1 mole of gas on the left-hand side, and none on the right-hand side. The right-hand side has the smaller volume. You can encourage a smaller volume by increasing the pressure. This will move the equilibrium to the **right**.

In (b), you are adding a **product**. This will encourage the reverse reaction. The equilibrium will move to the **left**.

(c) is the most difficult. If you add sodium hydroxide, the hydroxide ions will combine with the hydrogen ions to form water. This removes a product, so the reverse reaction is unable to take place. The forward reaction can still take place, so the equilibrium moves to the **right**.

Acids and Bases

Summary

- The pH scale runs from below zero to above fourteen.
- pH values below seven are acidic and those above seven are alkaline.
- pH seven is neutral.

The concentrations of H^+ ions and OH^- ions are linked by the relationship:

$$[H^+] \times [OH^-] = 10^{-14}.$$

If the concentration of H^+ ions is 10^{-x}, the pH is *x*.

For example, if the H^+ concentration is 10^{-2}, the pH is 2.

Example 8

What is the pH of a $0{\cdot}01\ mol\ \ell^{-1}$ solution of sodium hydroxide?

Solution

You can assume that since the sodium hydroxide, NaOH, is a solution of a strong alkali, it has broken down completely into ions. In this case, $[OH^-]$ is 10^{-2}.

In this case, the $[H^+] = 10^{-14} \div 10^{-2} = 10^{-12}$ so the pH is **12**.

Weak and strong acids

You can show acids by the formula HA, where A could be chloride, as in HCl, or ethanoate, as in CH_3COOH.

A strong acid (e.g. hydrochloric acid) breaks down almost fully into ions.

The equilibrium

$$HA \rightleftharpoons H^+ + A^-$$

lies to the right.

A weak acid (e.g ethanoic acid) produces very few ions.

The equilibrium

$$HA \rightleftharpoons H^+ + A^-$$

lies to the left.

Alkalis are similar. A weak alkali gives very few OH^- ions, and a strong one gives lots.

Behaviour of weak and strong acids

Since a strong acid contains lots of H^+ ions it:

- has a lower pH than the weak acid
- is a much better electrical conductor
- reacts faster than weak acids.

But equal volumes of weak and strong acids with the same concentration need just the same number of moles of OH^- ions to neutralise them.

Ethanoic acid, CH_3COOH, ionises as follows:

$$CH_3COOH \rightleftharpoons H^+ + CH_3COO^-$$

Just a few ions form. As soon as you add OH^- ions, they remove H^+. The reverse reaction stops but the forward reaction continues replacing the H^+. Adding more alkali removes more H^+ which are constantly replaced until all the acid has ionised and all the H^+ have reacted.

Weak alkalis compare with strong alkalis in just the same way as weak and strong acids.

Salts of weak and strong acids and alkalis

The salts of weak acids and strong alkalis are alkaline. The salts of strong acids and weak alkalis are acidic.

Sodium ethanoate is the salt of the weak acid, ethanoic acid, and the strong alkali, sodium hydroxide. When it is added to water, it dissolves and separates into ions.

$$Na^+CH_3COO^- \rightarrow Na^+ + CH_3COO^-$$

The ethanoate ion, being the ion from a weak acid, tries to become stable by forming ethanoic acid.

It does this by reacting with water as follows:

$$CH_3COO^- + H_2O \rightarrow CH_3COOH + OH^-$$

The hydroxide ions produced make the solution alkaline.

In a similar way, a solution of ammonium chloride in water is acidic. Ammonium chloride is a salt of the weak alkali, ammonium hydroxide and the strong acid, hydrochloric acid.

When it is added to water it breaks down as follows.

$$NH_4Cl \rightarrow NH_4^+ + Cl^-$$

The ammonium ion, being the ion from a weak alkali, tries to become stable by forming ammonium hydroxide.

It does this by reacting with water as follows:

$$NH_4^+ + H_2O \rightarrow NH_4OH + H^+$$

The hydrogen ions produced make the solution acidic.

Just remember that 'the strong one wins' since a salt of a weak acid and a strong alkali is alkaline, and the salt of a strong acid and a weak alkali is acidic.

Example 9

a) Sulphurous acid is a weak acid.
   What is meant by a weak acid?

b) Sodium sulphite is the salt of a weak acid (sulphurous acid) and a strong alkali. What happens to the pH of water when sodium sulphite is dissolved in it?

Solution

a) It is an acid which is not fully dissociated (not fully broken down into ions).

b) It rises above pH 7. (Remember, 'the strong one wins'.)

Example 10

Potassium cyanide, KCN, is made by the reaction of an acid with an alkali.

When dissolved in water, the solution has a pH of 8·5.

a) What do we learn about the strengths of the acid and the alkali involved?

b) Write the formula for the acid involved in this reaction.

Solution

a) Since the salt solution has a pH of 8·5, it is alkaline. Remember, 'the strong one wins'. The alkali must be **strong**, therefore the acid must be **weak**.

b) The formula is KCN. Since potassium is in group 1, it has a valency of 1. In which case the other ion, CN^-, (cyanide) must also have a valency of 1. Therefore the formula of the acid must be **HCN**.

Example 11

A solution of sodium ethanoate in water has a pH of 8.

Explain this observation.

Solution

Assuming that you understand that pH 8 is an alkaline solution, you can start off by explaining that this salt must be made from a weak acid and a strong alkali. That's usually enough to earn 1 mark. But you have to go on and explain that when the salt is added to water, the ethanoate ions produced react with the water as follows:

$$CH_3COO^- + H_2O \rightarrow CH_3COOH + OH^-$$

Redox Reactions

Summary

You met redox reactions at Standard Grade. You probably learned something like 'OILRIG' – 'oxidation is loss, reduction is gain' (of electrons). Redox is when both reactions happen.

Oxidising and reducing agents

An oxidising agent oxidises something else. It makes something else lose electrons. It gains these electrons – it gets reduced.

A reducing agent reduces something else. It gives it electrons. So the reducing agent loses electrons – it's oxidised.

Ion–electron equations

Make sure you can recognise oxidation and reduction. If electrons appear with a '+' sign in front of them on the **left-hand** side of the equation, it's reduction. Otherwise, it's oxidation.

Suppose you're asked to work out the ion–electron equation for the conversion of permanganate (MnO_4^-) to the Mn^{2+} ion. There's a standard routine:

- Balance all elements except hydrogen and oxygen.
- Balance the oxygen by adding water molecules as required to the side short of oxygen.
- Balance the hydrogen by adding hydrogen ions to the side short of hydrogen.
- Balance charge using electrons.

Let's apply this to the permanganate equation.

$$MnO_4^- \longrightarrow Mn^{2+}$$

The Mn is already balanced.

Balance the oxygen.

$$MnO_4^- \longrightarrow Mn^{2+} + 4H_2O$$

Balance the hydrogen.

$$MnO_4^- + 8H^+ \longrightarrow Mn^{2+} + 4H_2O$$

Add five electrons to the left to equalise the charge.

$$MnO_4^- + 8H^+ + 5e^- \longrightarrow Mn^{2+} + 4H_2O$$

Of course, in an actual example in an exam, you don't rewrite the equation step by step. You just build it up as you go.

Example 12

In a reaction, chlorate ions, ClO_3^- are reduced to chlorine.

$$ClO_3^- \rightarrow Cl_2$$

Complete the above to form the ion–electron equation.

Solution

Step 1: Balance all elements except hydrogen and oxygen. You need to double the number of chlorate ions.

$$2ClO_3^- \rightarrow Cl_2$$

Step 2: Balance the oxygen atoms using water molecules.

$$2ClO_3^- \rightarrow Cl_2 + 6H_2O$$

Step 3: Balance the hydrogen atoms using hydrogen ions.

$$2ClO_3^- + 12H^+ \rightarrow Cl_2 + 6H_2O$$

Step 4: Balance the charge using electrons.

$$2ClO_3^- + 12H^+ + 10e^- \rightarrow Cl_2 + 6H_2O$$

Of course, in an actual example in an exam, you don't have to keep rewriting the equation step by step. You just build up the starting equation as you go.

Example 13

Example 12 was quite easy. You can meet quite puzzling examples. Look at this one.

In the reaction of ethanal with Tollens' reagent (silver ions in ammonia), silver ions are reduced to metallic silver.

Complete the following ion–electron equation for the oxidation.

$$CH_3CHO \rightarrow CH_3COOH$$

Solution

This doesn't seem to fit in with the examples above. But if you examine the equation, and think about it, you can still apply the rules. The only thing about it is that there is one more oxygen on the right-hand side. So add a water molecule to the left-hand side and see what happens.

$$CH_3CHO + H_2O \rightarrow CH_3COOH$$

Now we've got two extra hydrogen atoms on the left-hand side, so add two hydrogen ions to the right-hand side.

$$CH_3CHO + H_2O \rightarrow CH_3COOH + 2H^+$$

Finally, you need two electrons on the right to balance the charge.

$$CH_3CHO + H_2O \rightarrow CH_3COOH + 2H^+ + 2e^-$$

You can see that the electrons end up on the right-hand side – so it really is an oxidation!

Example 14

Compounds of silver are used in black and white photography. Hydroquinone, $C_6H_6O_2$, is used as a developer. It reacts with silver ions as follows.

$$2Ag^+ + C_6H_6O_2 \rightarrow C_6H_4O_2 + 2Ag + 2H^+.$$

Write the ion–electron equation for the oxidation reaction.

Solution

It's not immediately clear how to do this. However, if you want the oxidation reaction, you don't want the reduction reaction. It **is** clear what the reduction reaction is.

$2Ag^+ \rightarrow 2Ag$ (silver ions have to gain electrons for this to take place)

So you have to pull this out of the equation, leaving:

$$C_6H_6O_2 \rightarrow C_6H_4O_2 + 2H^+$$

Everything here is balanced, except the charge. You can balance that by adding two electrons to the right-hand side, giving:

$$C_6H_6O_2 \rightarrow C_6H_4O_2 + 2H^+ + 2e^-$$

We'll look at redox calculations in the 'Calculations' chapter!

Electrolysis

We'll look at electrolysis calculations in the 'Calculations' chapter!

Nuclear Chemistry

Summary

Make sure you know this basic information.

Particle	Mass	Charge	Location
Proton	1 amu	+	Nucleus
Neutron	1 amu	0	Nucleus
Electron	Almost 0	–	Around nucleus

Summary *continued* ➢

Summary continued

Atomic number = number of protons

Mass number = number of protons + number of neutrons

Here's the nuclide notation for an atom of carbon.

$$^{13}_{6}C$$

This tells you that the atomic number is 6. The mass number is 13, so there must be 7 neutrons.

If the number of protons in the nucleus is balanced by a suitable number of neutrons, the atom will be stable. If there's an imbalance, the atom won't be stable, and will emit alpha, beta or gamma radiation till it is. Here's a summary of the features of these radiations.

Radiation	What it is	Effect on atomic number	Effect on mass number	Range in air
Alpha, α	2 protons + 2 neutrons	Lowers it by two	Lowers it by four	centimetres
Beta, β	Electron	Raises it by one	No effect	metres
Gamma, γ	Energy	No effect	No effect	kilometres

Half-life

If you have a collection of radioactive atoms, they will all eventually undergo radioactive decay. You can't predict when a given atom will do this but, after a certain time, half the atoms in the collection will have decayed. This time is called the 'half-life' of the isotope, and it's constant. It doesn't matter whether you start with a billion of the atoms or just ten. It takes the same time for a billion of them to change to half a billion as it does for ten to change to five!

Figure 4.4 Does a radioactive cat have 18 half lives?

Applications of radioisotopes

Cobalt-60 is used to treat cancer. Its gamma rays have high penetrating power and can be focussed on the tumour.

Iodine-133 is used to treat conditions of the thyroid gland in the neck. If the gland is enlarged, iodine-133 can be taken into the body. The iodine gathers in the gland and will destroy part of it.

Cobalt-60 is also used to monitor the thickness of a product which is rolled out in thin strips. The amount of radioactivity which passes through the strip depends on the thickness.

Information about the intensity of radiation passing through the strip is sent to a computer which then adjusts the spacing of the rollers to ensure an even thickness.

If you're asked to select an isotope for a particular purpose, think about the half-life and the type of radiation. If it's an industrial process, you should pick an isotope with a **long** half-life so that it doesn't have to be replaced frequently. If the isotope is going to be used in a human body, pick a **short** (but not too short) half-life so that the radioactivity isn't there for longer than needed. You wouldn't put an alpha emitter into the body, because the radiation is extremely damaging (and most alpha emitters are very toxic anyway!).

Carbon dating

The atmosphere contains some carbon dioxide made from the radioactive isotope, carbon-14. Through photosynthesis, this radioactive carbon dioxide enters plants and then animals. While the plants and animals are alive, the concentration of radioactivity stays constant. When the organism dies the amount of radioactivity falls steadily. The half-life of carbon-14 is 5730 years. If the radioactivity of an ancient wooden object is half of that in a living organism, the object is about 5730 years old. If it's a quarter, it's 11 460 years old. However, the method can't be used for objects more than about 50 000 years old, because the level of radioactivity is too low.

Figure 4.5 We're all a bit radioactive

Fusion and fission

In fission, atoms are bombarded with neutrons. This makes them split into smaller bits, releasing neutrons and energy as they do so. These neutrons split other atoms, releasing energy and yet more neutrons. In this way a vast amount of energy is released in a very short time. This process is used in nuclear power stations and is also the basis of nuclear weapons.

In fusion, smaller atoms join together, releasing a lot of energy. This takes place in stars, and is how all the elements were formed.

Figure 4.6 Lots of fusion goes on here

Example 15

Radioactive strontium would differ from its stable isotopes in:

A atomic mass

B chemical properties

C atomic number

D electronic configuration.

Solution

Let's deal with the wrong answers first. What makes an atom radioactive is its proton to neutron ratio. It's nothing to do with electrons. So it's **not** D. The chemical properties depend on the electrons, so it's **not** B either. Atoms of strontium all have the **same** atomic number so it's **not** C. It must be A.

Remember from Standard Grade – isotopes are atoms with the <u>same</u> atomic number but <u>different</u> mass numbers.

Example 16

A radioactive isotope used in a hospital has a half-life of 2 hours. At midday it has a count rate of 12 000 counts per minute.

a) What would the count rate be at 6p.m. that day?

b) A compound of the isotope was dissolved in water.

What effect would this have on the half-life of the isotope?

c) Give a use for radioactive isotopes in medicine.

Solution

a) From midday to 6p.m. is 6 hours. Since the half-life is 2 hours, this means that three half-lives will pass. After one half-life, the intensity of radiation is $\frac{1}{2}$ of its original level. After two half-lives, it's $\frac{1}{4}$. After three half-lives, it's $\frac{1}{8}$. So the count rate will be $\frac{1}{8}$ of 12 000, that is 1500 counts per minute.

b) Remember, nothing can affect the half-life – no effect!

c) See the examples above!

Example 17

Sodium-24 is a radioactive isotope.

a) What effect will a temperature increase have on the rate of radioactive decay?

b) How will the intensity of radiation from 1 g of sodium-24 compare with 1 g of sodium chloride made from sodium-24?

Solution

a) The rate of radioactive decay is controlled by the half-life. You can't alter the half-life so the answer is – no effect!

b) The intensity of radiation isn't the same as the half-life. If you think about it, the 1 g of sodium chloride is made up of sodium and chlorine – it isn't pure sodium. So compared with the 1 g of sodium, there's less sodium, so the intensity of radiation has to be **less**.

Example 18

A diagram of a smoke detector is shown.

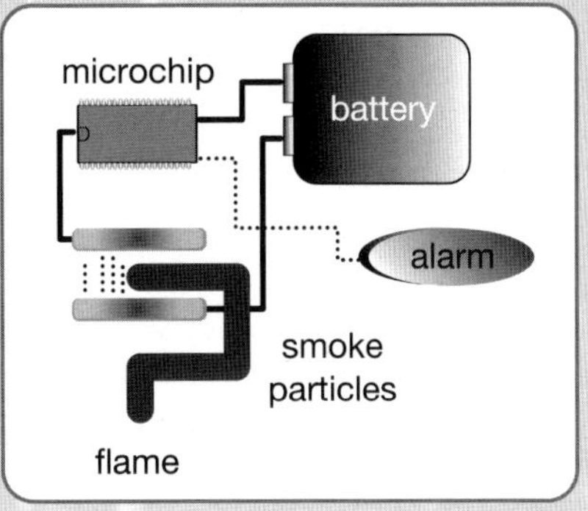

Some smoke detectors use the alpha radiation from americium-241 (half-life 432 years).

Give **two** reasons why americium-241 is suitable for this application.

Solution

One reason is that the isotope used has a long half-life. The radioactivity isn't going to dwindle away so the device will keep on working (if you make sure the battery is working too!). The other reason is that alpha particles can travel only a few centimetres in air. Up on the ceiling, there's not much chance that the detector can do us any harm. The radiation can't even get out of the smoke detector!

Chapter 5

PROBLEM SOLVING

When you did your Standard Grade exam you could tell which questions were Problem Solving because the marks for the question were shown in the Problem Solving column down the side of the page. However, in Higher, the problem solving marks are not shown separately, so you can't necessarily tell from looking at the question whether it is Knowledge and Understanding or Problem Solving. One thing you can be sure of is that there won't be any 'easy' problem solving marks for questions like drawing a line graph, or making a table, or spotting a simple trend in some data.

The Higher Problem Solving questions are more difficult and often draw on aspects of chemistry and chemical techniques you'll never have heard of. When you're faced with a question of this type you have to read it carefully a couple of times to make yourself quite clear about what the question is about. Since it probably involves something you have never met, it will start off by setting the scene and this may take several lines. You must read it all carefully. Then try to be clear about what you are being asked to do. If you have read the introduction carefully, you should manage the questions without too much difficulty. Often, these questions have nothing at all to do with the content of the Higher Chemistry course but are completely self-contained. They are included because they meet the requirements for different kinds of problem solving which are included in the Arrangements for Higher Chemistry. However, a lot of questions which are mainly Problem Solving have marks for Knowledge and Understanding as well, so it's unlikely that any long question will be purely Problem Solving.

There are various kinds of Problem Solving questions, and this chapter shows you examples of each.

Apparatus Questions

A fairly common kind of problem solving question asks you to complete a drawing of chemical apparatus.

Example 1

A student added hydrochloric acid to magnesium ribbon to produce hydrogen.

The hydrogen produced in the reaction can be contaminated with small amounts of hydrogen chloride vapour. The vapour is very soluble in water.

Complete the diagram (overleaf) to show how the hydrogen chloride can be removed before the hydrogen is collected.

***Example** continued* ➢

Example 1 continued

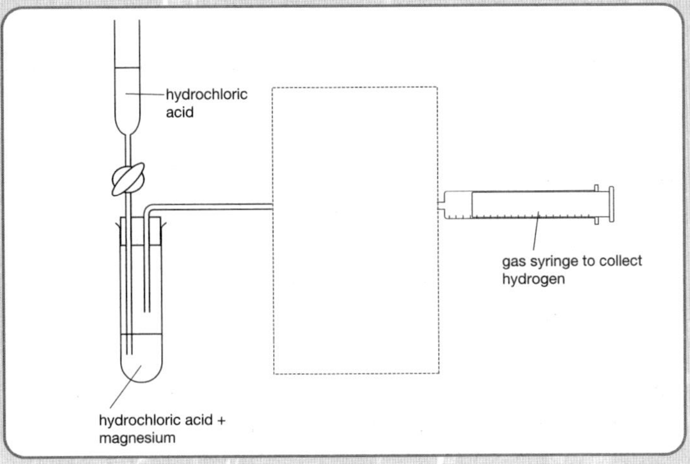

Solution

The important thing to realise is that you're <u>not</u> being asked to invent any <u>new</u> apparatus. It will involve apparatus you should have seen many times in the lab at school.

The method simply involves allowing the gases to pass through water, which will dissolve the hydrogen chloride. You're obviously expected to know that hydrogen itself does not dissolve in water. That's something you should have learned as far back as S2, when you probably made hydrogen and collected it over water.

So all you need to draw is shown below.

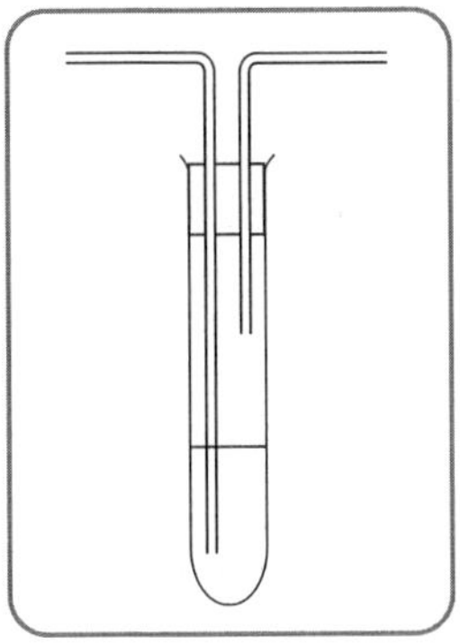

It's really important that what you draw will actually work. The tubes in the test tube have to be the right length. Look at some of the ways you could go wrong!

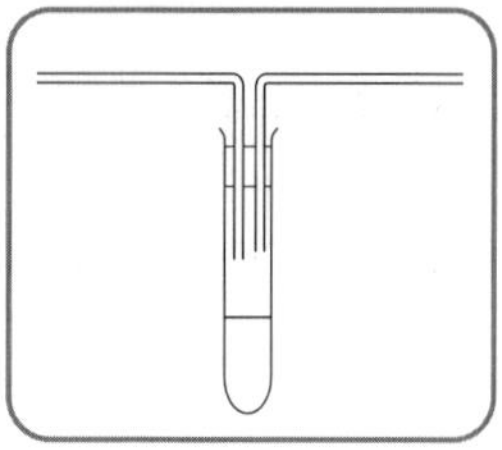

In this diagram, the inlet tube, on the left is too short. This means that the gases don't have the chance of passing through the water, so the hydrogen chloride will not be removed.

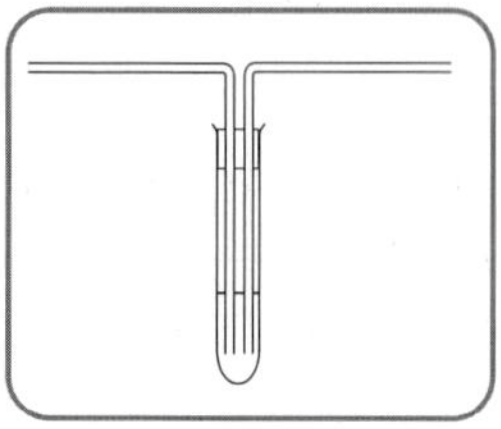

In this diagram, the outlet tube (on the right) is too long. This means that, as gases enter the tube, water is forced up the outlet tube and will be collected in the syringe, instead of hydrogen. You could also make use of a test tube with side arm, and that gets rid of the problem very neatly.

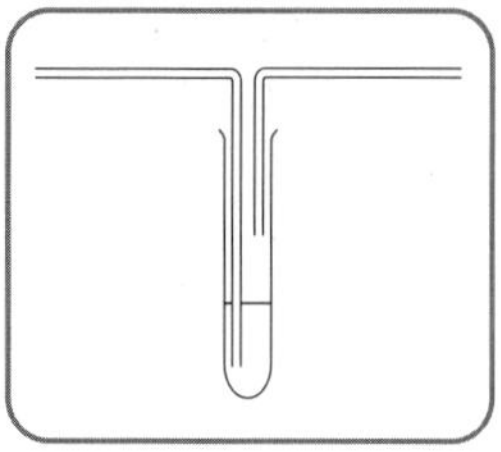

In this diagram, something important is missing. You've spotted it! There's no stopper in the test tube. The gases are just going to escape into the air. It can't possibly work.

Example 2

In some countries, cow dung is fermented and the mixture of gases produced, known as biogas, is used as a fuel. The mixture contains a small amount of carbon dioxide.

a) Name the main component of the biogas mixture.

b) The percentage of carbon dioxide in a biogas sample can be found by experiment. Part of the apparatus required is shown in the diagram.

Complete the diagram to show all of the apparatus which could be used to carry out the experiment.

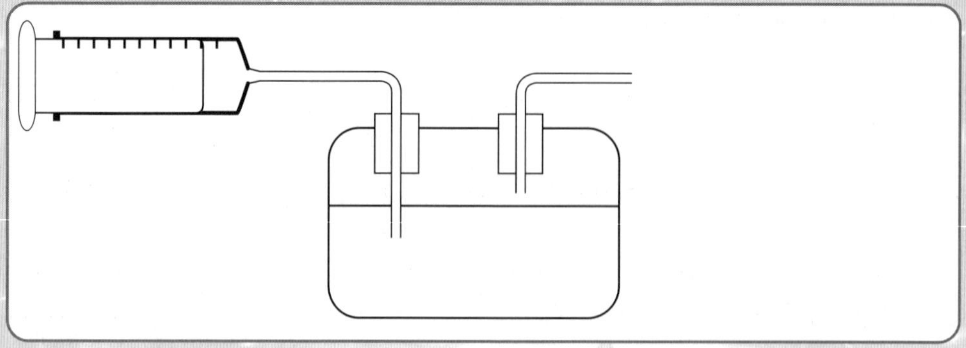

Solution

a) This is simply knowledge and understanding. The main component of biogas is methane.

b) All you have to do here is add another gas syringe on the right, like the one on the left, to collect the methane.

But watch out. As in the example above, you must be sure not to close off any tube which should be open.

Note also that you have to know how much methane you have collected. The gas syringe on the left has a scale on it. The one on the right must also have a scale, otherwise you will not know how much gas you have collected, and you may not collect full marks for the question either!

Another way of measuring the volume of methane would be to collect the gas over water in an inverted measuring cylinder. However, you must take care that your diagram does not contain the common error of showing the delivery tube passing through the side of the water container or the measuring cylinder. It must pass over the rim of the water container and under the mouth of the measuring cylinder as shown in the left hand diagram.

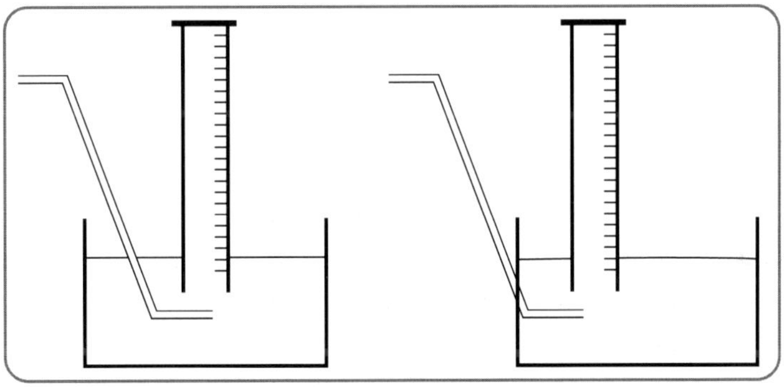

Which is Which?

Another kind of problem solving question asks you to identify a test which will distinguish between two compounds.

Example 3

Dewar benzene is an isomer of benzene. It has the following structure.

Describe a chemical test which could be used to distinguish between benzene and Dewar benzene.

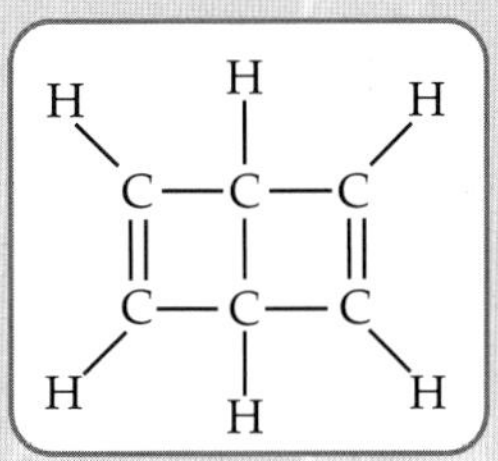

Figure 5.1 Structure of Dewar benzene

Solution

Although benzene only appears in Higher papers, this is really drawing on your knowledge of Standard Grade chemistry. The answer simply involves using the bromine water test. Bromine does not react readily with benzene, however, it will react with Dewar benzene. This is because it has two carbon-to-carbon double bonds.

Chemical tests

It is a good idea to be thoroughly familiar with as many chemical tests as possible, since these will help you to tackle this kind of problem solving question.

What you should know about chemical tests

Here is a list of tests. You should learn it. You could even make up a set of flash cards and test yourself.

Test	Distinguishes
Bromine solution	Alkenes and alkynes will react with bromine, decolourising it. Alkanes and benzene do not react readily.
Benedict's Reagent	Gives a positive result for reducing sugars, i.e. glucose, fructose and maltose. Negative for starch and sucrose.
Starch	Iodine (gives blue–black colour).
Iodine	Starch (gives blue–black colour).
Fehling's solution	Gives a positive result for aldehydes and a negative result for ketones.
Tollens' reagent	Gives a positive result (silver mirror) for aldehydes and a negative result for ketones.
Acidified potassium dichromate	Reacts with primary and secondary alcohols, but not with tertiary alcohols.
Heating compound with alkali releases ammonia	Ammonium compound.
Addition of acid releases carbon dioxide	Carbonate.
Orange-yellow flame test	Sodium compound.
Lilac flame test	Potassium compound.
Gas which burns with a 'pop'	Hydrogen.
Gas which turns lime water cloudy	Carbon dioxide.
Gas which relights a glowing taper	Oxygen.

Flow diagrams

Questions based on flow diagrams, often describing industrial processes, have always been popular in Higher Chemistry exams. This is partly because it is possible to ask a number of questions based on the same diagram. It also allows you to focus in on the question thus cutting down the number of changes of theme you have to make during the exam. The Higher course contains a topic on the Chemical Industry, and this makes it likely that questions like this will turn up from time to time.

Example 4

The flow diagram in Figure 5.2 shows the production of sodium carbonate by the Solvay Process.

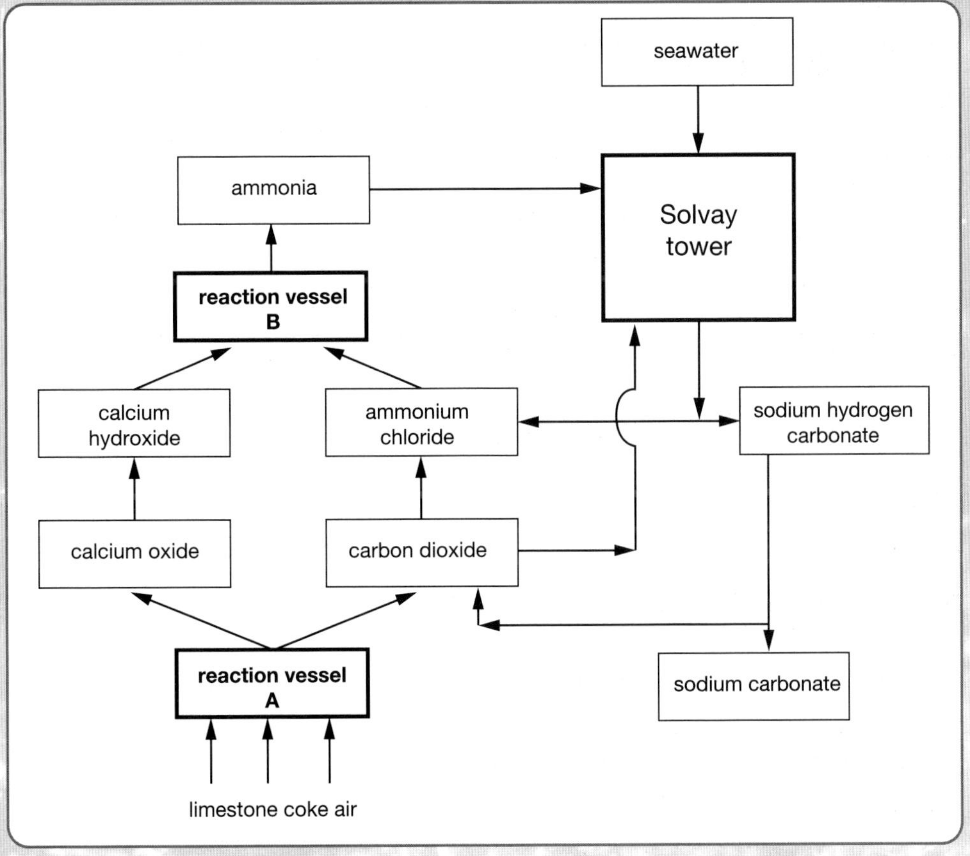

Figure 5.2 The Solvay process

***Example** continued* ➢

Example 4 continued

a) Name the reactants in the reaction taking place in the Solvay tower.

b) In reaction vessel A, carbon dioxide is produced by the following two reactions.

$$CaCO_3(s) \rightarrow CaO(s) + CO_2(g) \qquad \Delta H = \_\_\_\_\_\_$$

$$C(s) + O_2(g) \rightarrow CO_2(g) \qquad \Delta H = \_\_\_\_\_\_$$

For each reaction, add a sign after the ΔH to show whether the reaction is endothermic or exothermic.

c) As well as ammonia, a salt and water are produced in reaction vessel B.

Write a balanced equation for the production of ammonia in this reaction vessel.

d) The seawater used in the Solvay process can contain contaminant magnesium ions. These can be removed by the addition of sodium carbonate solution.

Why is sodium carbonate solution suitable for removing contaminant magnesium ions?

e) Using the information in the flow diagram, give two different features of the Solvay process that make it economical.

Solution

The first thing you should do, when faced with a question like this is to inspect it closely, trying to understand what is going on, without even looking at the questions based on the diagram. Try to understand the chemistry – almost certainly, it will be simple, maybe based on Standard Grade. (In any case, you don't meet many new chemical reactions in Higher, apart from the chemistry of carbon in Unit 2.) Besides, the chemical industry likes reactions which are simple, because they are usually economical to carry out.

Your first problem with this question is to understand where to start. You should notice that reaction vessel A is shown in bold, and limestone (what's that?), coke (what's that?) and air go into it. Suppose you don't know what limestone is or what coke is (although if you have revised your Standard Grade work thoroughly, you will!) – can you do the question? Of course you can! The products of the reaction are carbon dioxide and calcium oxide. So it's a fair bet that limestone is calcium carbonate and coke is carbon.

If you follow the chemistry up to reaction vessel B you will find ammonium chloride reacting with calcium hydroxide, to form ammonia. (In Standard Grade you met the idea that any hydroxide compound reacting with any ammonium compound would produce ammonia.)

The key to the whole thing is the Solvay tower. In go ammonia, carbon dioxide and seawater. You can assume that seawater is a solution of sodium chloride. Out come ammonium chloride (which goes into reaction vessel B) and sodium hydrogen carbonate. The sodium hydrogen carbonate ends up as sodium carbonate (the end product) and carbon dioxide. You will notice that the carbon dioxide ends up being fed back into the Solvay tower.

So, you've examined the diagram closely, and have got some idea about what's going on. Obviously, it's a good idea to have revised Standard Grade chemistry, because there are no chemical reactions from the Higher course in this question. That's not to say the question doesn't involve ideas from Higher, otherwise it wouldn't be in the exam.

By examining the diagram, you can already answer (a) – the reactants are ammonia, sodium chloride and carbon dioxide.

How about (b)? Which reaction would you expect to be exothermic – to give out heat?

$C(s) + O_2(g) \rightarrow CO_2(g)$ is just the burning of carbon. It's bound to give out heat, so ΔH has to be **negative**.

What about the other reaction? This involves the breakdown of $CaCO_3(s)$ (You were right! It was calcium carbonate!). Breaking a chemical down can usually be done by heating it, so ΔH has to be **positive**.

(c) is straight out of Standard Grade.

$$Ca(OH)_2 + 2NH_4Cl \rightarrow 2NH_3 + 2H_2O + CaCl_2$$

You have a lot to do here – work out five chemical formulae, and then balance the equation. All for just one mark. If you haven't balanced it, but all the formulae are correct, you're certainly in line for half a mark.

(d) depends on the fact that magnesium carbonate is not soluble in water, and is therefore removed by precipitation. You can check out the solubility of magnesium carbonate in the Data Booklet. Unfortunately, the question doesn't suggest that you look at the solubility table in the Data Booklet – you have to work this out for yourself!

(e) asks you to suggest features of this process that make it economical.

You could comment on the costs of the raw materials – limestone, coke, air and sea water. They are all inexpensive (although coke has to be made from coal, by heating it in the absence of air, and heat is always expensive).

You should also have noticed that a number of chemicals are **recycled** in the process – carbon dioxide from the sodium hydrogen carbonate, and ammonia, which is recycled as ammonium chloride. The more chemicals that can be recycled within the process, the more economical it will be.

On the whole, this is a difficult question but by carefully inspecting the diagram before you start to answer the questions, you will make it easier. You also have to be familiar with your Standard Grade work.

Specialised Problems

Some problem solving questions are built round investigative techniques which have nothing to do with what you have learned in school. They may reflect a specialism of one of the team who set the paper or even ongoing research which has had recent publicity.

Here is an example of a question based on research which one of the setting team had once done. It involves a technique called X-ray diffraction which is used to find molecular structures from electron density contours of molecules.

Example 5

X-ray diffraction is a technique used to determine the structures of molecules. It is the electrons in the atoms of the molecule which diffract the X-rays. From the diffraction pattern, an electron density contour map of the molecule can be constructed.

The map shown in Figure 5.3 was obtained using an aromatic compound with molecular formula $C_6H_3Cl_3O$.

a) Suggest why the hydrogen atoms do not show up clearly in the electron density contour map.

b) Draw the full structural formula for this compound.

c) Draw the electron density contour map that would be obtained for methanoic acid.

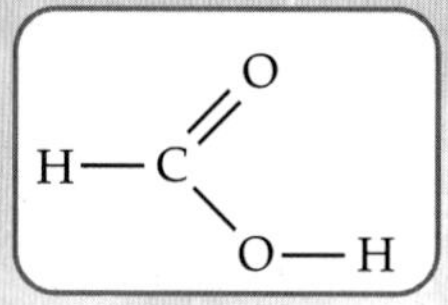

Figure 5.3 Electron density contour map

Solution

The question looks very difficult, but this is really only because it is very unfamiliar in appearance. You are told that the compound is aromatic, so you know that the compound is based on benzene, C_6H_6. In any case, the formula tells you that there are six carbon atoms, so the hexagonal shape in the middle must be made up of six carbon atoms.

a) There are three large atoms, and a smaller one, attached to the benzene ring. Are the three atoms the three hydrogens or the three chlorines? There's no need to worry about this, because part (a) of the question asks you why hydrogen atoms don't show up in such diagrams. The answer to this is that hydrogen atoms only have one electron so there isn't very much to show up in the diagram. The three big atoms must be chlorine, and the smaller one must be oxygen.

b) You were asked to draw the full structural formula for the compound.

So far, you have got to this point.

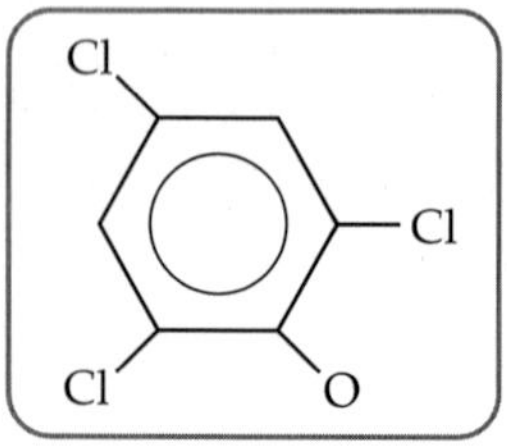

You still have to place the three hydrogen atoms. The only places they can go are on the two carbon atoms to which nothing is attached, and the oxygen, to give an OH group. So the final structure looks like this.

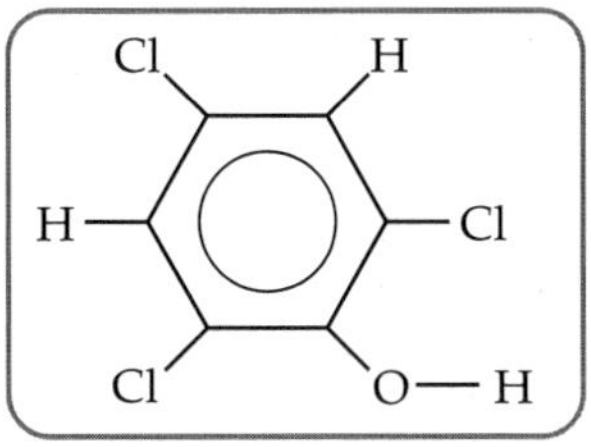

Since the question asks for a full structural formula, you should draw the hydroxyl group as shown, and not as –OH since that is not a full structural formula. In a full structural formula you must show all the bonds. In addition, you'd probably be wise to show the six carbon atoms, one at each corner, because a full structural formula would show all the atoms.

In fact, this compound is called trichlorophenol, or TCP for short – a well known antiseptic!

c) You were asked to draw the diagram you would expect for methanoic acid.

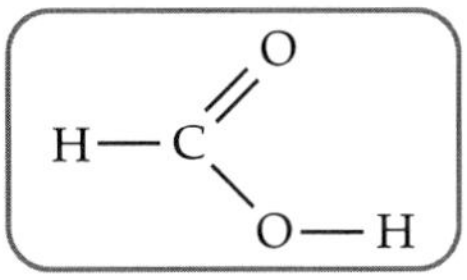

You have to remember that the two hydrogen atoms will not show up in the diagram, so all you have to do is show one carbon atom and two slightly larger oxygen atoms. Your diagram will be something like this.

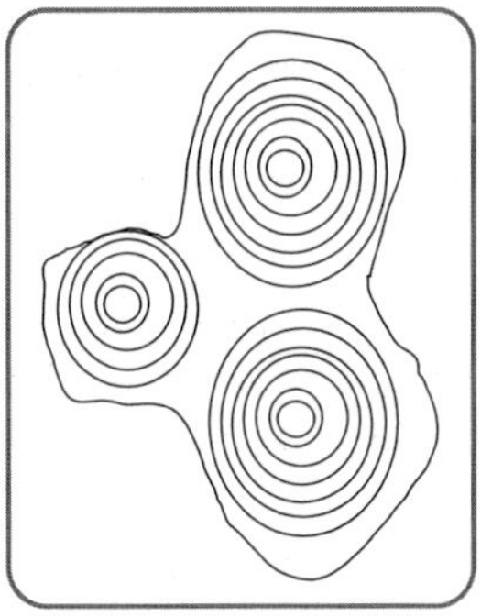

It's important that you show the carbon atom as being smaller than the two oxygen atoms.

The following question involves the technique of electrophoresis, again something that is not carried out in most schools. This technique involves the movement of ions in electric fields. You have met this idea in Standard Grade where you may have been shown an experiment where coloured ions are made to separate by applying a voltage. However, this question is about the identification of particular amino acids.

Example 6

Electrophoresis, widely used in medicine and forensic testing, involves the movement of ions in an electric field. The technique can be used to separate and identify amino acids produced by the breakdown of proteins.

a) Name the type of reaction that takes place during the breakdown of proteins.

b) The amino acid, glycine, has the following structural formula.

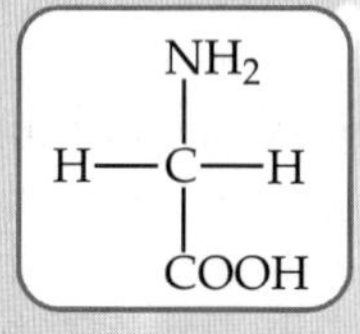

Like all amino acids, glycine exists as ions in solution and the charge on the ions depends on the pH of the solution. In solutions with low pH, glycine exists as a positively charged ion.

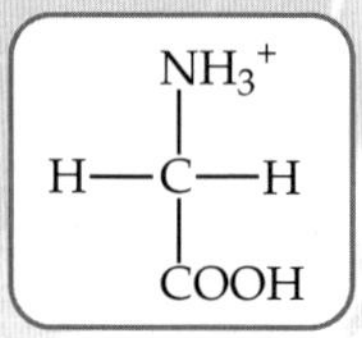

In solutions with a high pH, glycine exists as a negatively charged ion.

Draw the structure of this negatively charged ion.

c) The table overleaf shows the structures and molecular masses of three amino acids, A, B and C.

A mixture of amino acids, A, B and C, was applied to the centre of a strip of filter paper which had been soaked in a solution of pH 2. All three amino acids exist as ions in this acidic solution. A high voltage was then applied across the filter paper.

The amino acid ions separate according to their charge and molecular mass.

On the diagram shown in Figure 5.4, indicate the approximate positions of A, B and C once electrophoresis has separated the ions.

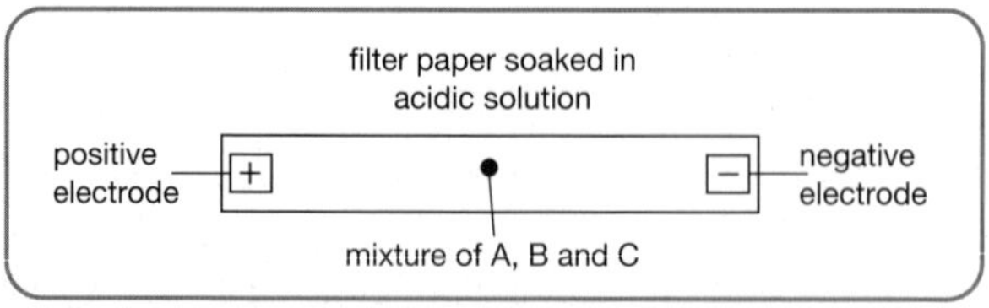

Figure 5.4 Apparatus for electrophoresis

Solution

a) This isn't Problem Solving – it's just a simple bit of Knowledge and Understanding. The breakdown of proteins to amino acids is hydrolysis.

b) This part of the question tells you that in solutions with a high pH (alkaline solutions), glycine exists as a **negatively** charged ion. You were asked to draw the structure of this ion. To answer this, you have to use your knowledge of how weak acids behave. At high pH, when there are lots of OH^- ions around, the COOH group will lose its H^+, to form water, and you are left with a COO^- (carboxylate) group. So the structure of the negative ion is as shown below.

$$
\begin{array}{c}
NH_2 \\
| \\
H-C-H \\
| \\
COO^-
\end{array}
$$

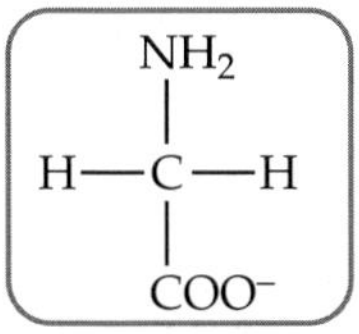

Amino acid	Structure	Molecular mass
A	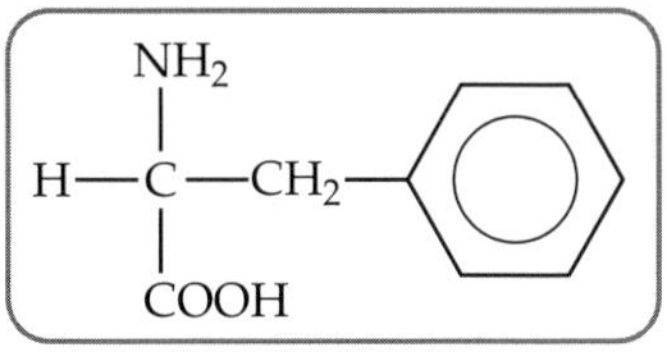 $\begin{array}{c} NH_2 \\ | \\ H-C-CH_2- \bigcirc \\ | \\ COOH \end{array}$	165
B	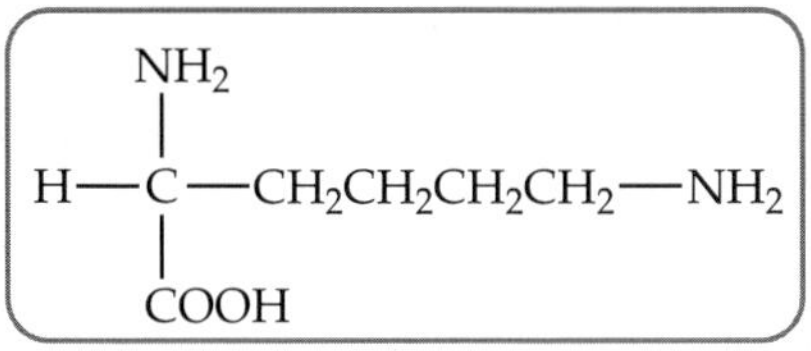 $\begin{array}{c} NH_2 \\ | \\ H-C-CH_2CH_2CH_2CH_2-NH_2 \\ | \\ COOH \end{array}$	146
C	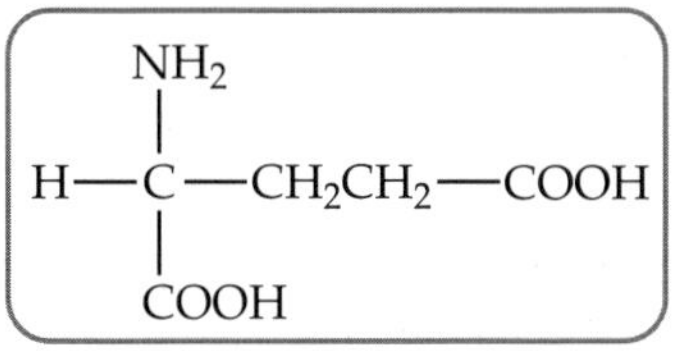 $\begin{array}{c} NH_2 \\ | \\ H-C-CH_2CH_2-COOH \\ | \\ COOH \end{array}$	147

c) The question goes on to show you the structures of three amino acids, A, B and C.

You were told that the amino acids separate according to their charge and molecular mass. On the diagram you are to show the approximate positions of A, B and C once the voltage has separated the ions.

This looks a tough question. However, the question told you earlier that in solutions of low pH , the $-NH_2$ group becomes $-NH_3^+$. This means that A will have a charge of 1+, B will have a charge of 2+ and C will have a charge of 1+. The bigger the charge, the further the ion will move. They must all move toward the negative electrode, since opposite charges attract.

You were also told that the ions separate according to their mass. A is much heavier than B or C, so it will not move so far. So you would expect A to move a little toward the negative electrode and B and C to move further toward the negative electrode. However, since B has a charge of 2+, and C only has a charge of 1+, you would expect B to move further than C.

You should show the positions as shown in Figure 5.5.

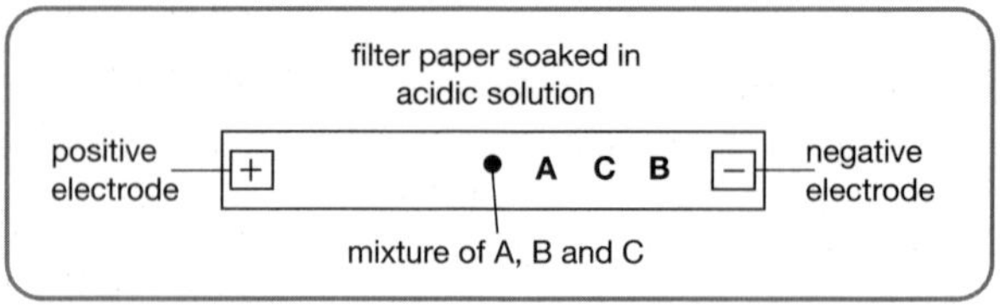

Figure 5.5 Apparatus for electrophoresis showing positions of amino ions

This question involves a lot of reading, and includes a number of new ideas. That is why it is so important to read it all very carefully before attempting to answer the questions.

Here are now some more examples that show the variety of problem solving questions.

Example 7

Perfumes normally contain three groups of components called the top note, the middle note and the end note. The top note compounds vaporise easily. They are very volatile. Two examples of such compounds are shown in Figure 5.6.

a) Explain why these compounds are likely to have pleasant smells.

b) Describe a chemical test which would allow you to distinguish between these two compounds. Give the result of the test.

c) The middle note compounds are less volatile (evaporate less easily). A typical middle note compound is shown in Figure 5.7 overleaf.

As a result of hydrogen bonding, this compound forms a vapour less easily than *p*-cresyl ethanoate.

Copy the molecule, and draw a second molecule of the same compound. Add a dotted line to show where a hydrogen bond could form between the two molecules.

d) The end note of a perfume has a long lasting fragrance. A typical end note compound is shown in Figure 5.8 overleaf.

Draw the structure of the alcohol which would be formed by reduction of civetone.

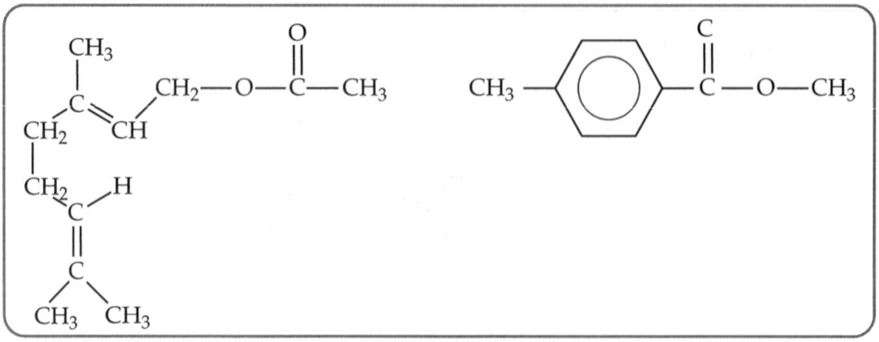

Figure 5.6 Geranyl ethanoate and *p*-cresyl ethanoate

Solution

This is a pretty difficult looking question. It is problem solving because of the unusual appearance of the molecules. However, if you take time to look at the molecules and think about them, it is really quite easy.

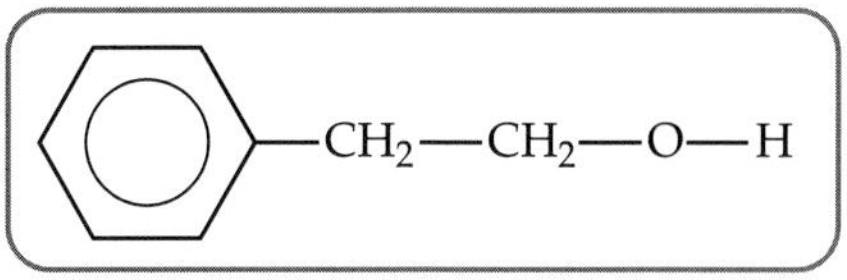

Figure 5.7 2-phenylethanol

a) The reference to pleasant smells should make you think of esters. Look at the molecules. You should spot the ester links without too much difficulty.

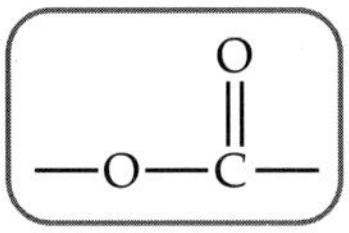

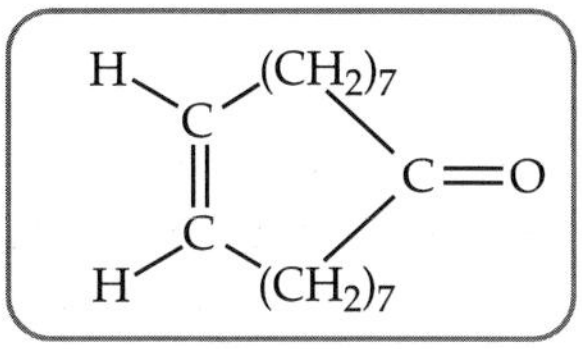

Figure 5.8 Civetone

In any case, the name 'ethanoate' should also be a clue that these are esters.

b) Notice that the geranyl ethanoate has two carbon to carbon double bonds. The other compound hasn't. All you have to do is carry out the bromine water test (from Standard Grade) and observe that geranyl ethanoate decolourises the solution. The other compound will have no effect.

Remember – aromatic compounds like *p*-cresyl ethanoate don't react readily with bromine water.

c) You should remember that hydrogen bonding takes place when there are H–F, H–O and H–N bonds around. So what you have to draw is something like this:

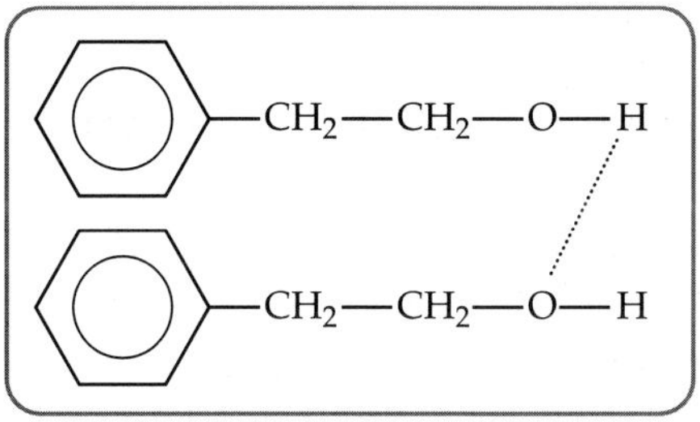

d) You need to be careful here. 'Reduction' in organic chemistry usually means adding hydrogen. Hydrogen can be added across the carbon-to-carbon double bond, and also to the carbonyl (C = O) group. A hydrogen atom adds at each end of the double bond, so an —OH (hydroxyl) group forms from the carbonyl group.

The product is this molecule.

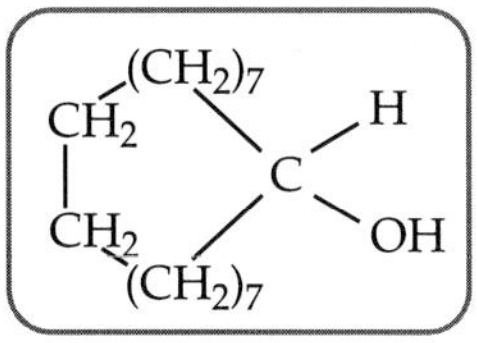

Example 8

A student used the reaction between magnesium ribbon and very dilute hydrochloric acid to find the molar volume of hydrogen gas.

The equation for the reaction is:

$$Mg + 2HCl \rightarrow MgCl_2 + H_2.$$

The following equipment and chemicals were used:

100 cm^3 measuring cylinder

250 cm^3 beaker

Filter funnel

Very dilute hydrochloric acid

Magnesium ribbon

a) Draw a diagram to show how the student would have set up the above apparatus and chemicals to find the molar volume of hydrogen.

b) What measurements would be necessary?

c) How would the measurements be used to calculate the molar volume of hydrogen?

Solution

a)

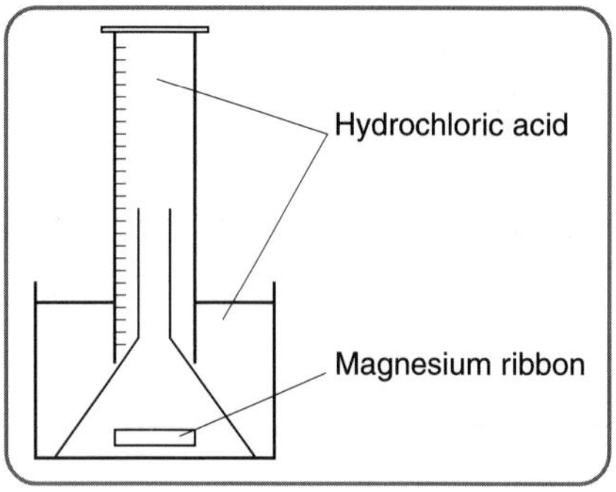

The idea of this apparatus is that magnesium reacts with hydrochloric acid to produce hydrogen. The bubbles of hydrogen are channelled into the measuring cylinder, where the volume of gas can be measured.

b) The molar volume of a gas is the volume occupied by one mole of the gas.

You need to find a link between the volume of gas collected and the number of moles. Magnesium is used up in the reaction. The equation tells you that the moles of hydrogen are the same as the moles of magnesium. If you weigh the magnesium at the start and after you have collected a big enough volume of hydrogen to measure accurately, you can find the mass of magnesium which has reacted, and hence the number of moles.

c) The moles of magnesium reacted equal the moles of hydrogen formed.

By dividing the volume of hydrogen by the number of moles of hydrogen, the volume per mole is found. This is the molar volume.

Example 9

A proton NMR spectrum can be used to help identify the structure of an organic compound.

The three key principles used in identifying a group containing hydrogen atoms in a molecule are as follows.

1. The position of the line(s) on the *x*-axis of the spectrum is a measure of the 'chemical shift' of the hydrogen atoms in the particular group.

   Some common 'chemical shift' values are given in the table below.
2. The number of lines for the hydrogen atoms in the group is $n + 1$ where n is the number of hydrogen atoms on the carbon atom next to the group.
3. The maximum height of the line(s) for the hydrogen atoms in the group is relative to the number of hydrogen atoms in the group.

The spectrum for ethanal is shown in Figure 5.9.

a) The chemical shift values shown in the table are based on the range of values shown in the Data Booklet for proton NMR spectra.

   Use the Data Booklet to find the range in the chemical shift values for hydrogen atoms in the following environment.

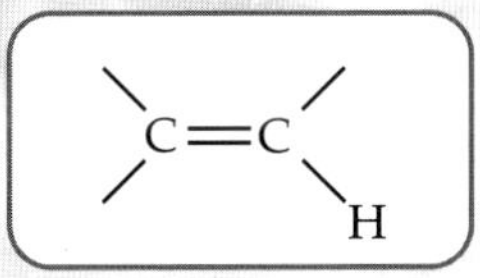

b) A carbon compound has the spectrum shown in Figure 5.10.

   Name this compound.

c) Using Figure 5.11, draw the spectrum that would be obtained for chloroethane.

Group containing hydrogen atoms	Chemical shift
$-CH_3$	1.0
$-C \equiv CH$	2.7
$-CH_2Cl$	3.7
$-CHO$	9.0

Solution

Help! How do you start this – you don't know anything about NMR!

Don't worry – you're not expected to! But you are expected to read the information thoroughly and think about it.

The first key point simply tells you that where the lines appear on the *x*-axis indicates the structure of the group which contains the hydrogen atoms. If it appears at 1, then it's a $-CH_3$ group, and so on.

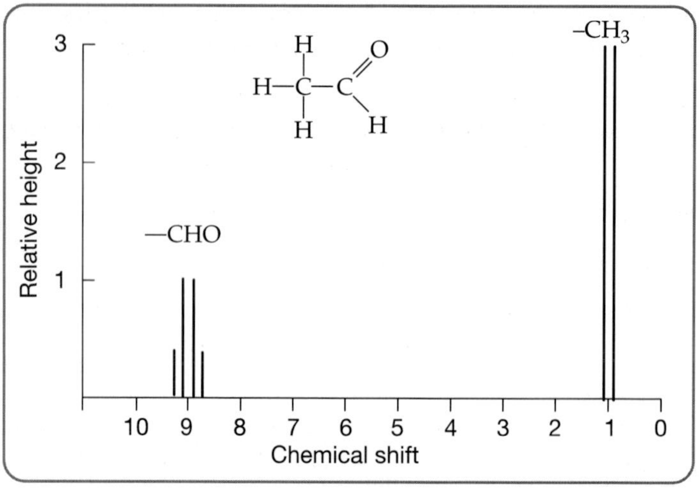

Figure 5.9 Proton NMR spectrum for ethanal

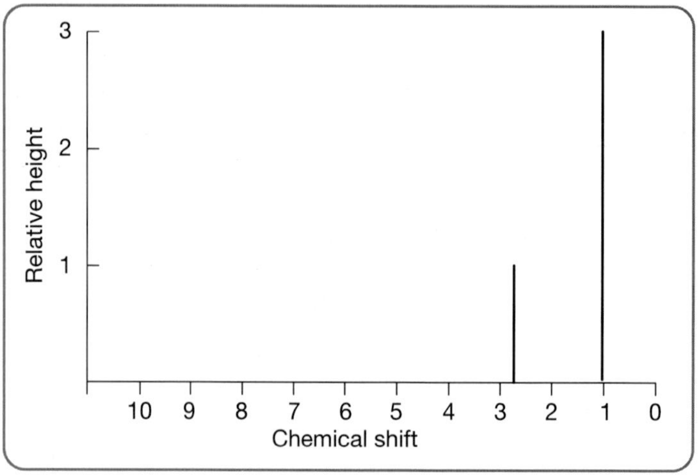

Figure 5.10 Proton NMR spectrum for unknown carbon compound

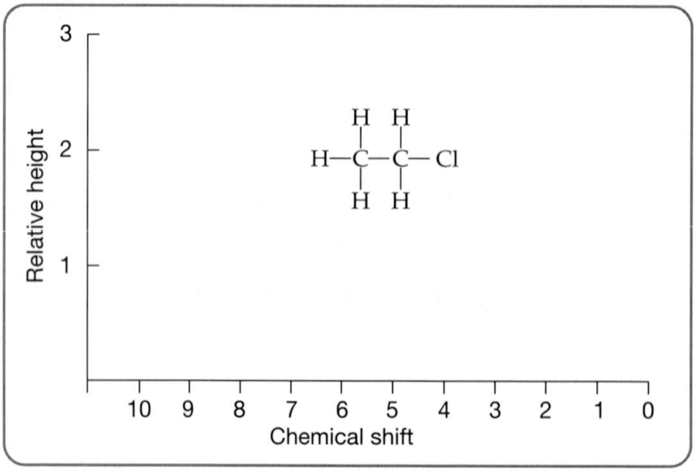

Figure 5.11

Check that the example spectrum matches the information in the table. That is, do the lines for $-CH_3$ and $-CHO$ appear at the correct positions on the horizontal axis?

Read the second key point again. The $-CHO$ group gives four lines. That's one more than the number of Hs on $-CH_3$. The $-CH_3$ group gives two lines. That's one more than the number of Hs on $-CHO$. It all fits in with what you've been told.

Read the third key point again. The height is a measure of the number of carbon atoms.

Now you're ready to look at the actual questions.

a) This is an easy starter – you just have to look up the Data Booklet.

It's page 15. (That's a page you haven't looked at before!)

The answer is **5.5–4.5**.

b) Single lines? If $1 = n + 1$, then $n = 0$. The neighbouring carbon atoms must have no hydrogen atoms on them. There's obviously a $-CH_3$ group. The line at 2.7 is caused by a $-C \equiv CH$ group. Can the compound have this structure?

$$CH_3 - C \equiv CH$$

Watch out, though. You have to name the compound. It's **propyne**.

(If you gave the structure, rather than the name, you'd probably get a $\frac{1}{2}$ mark.)

c) This compound has a $-CH_3$ group, so the line will be at 1 and will have a height of 3. The neighbouring carbon atom has two hydrogens, so there should be three lines $(n + 1)$.

The rest of the molecule, $-CH_2Cl$, will give a line at 3.7. The neighbouring carbon atom (the $-CH_3$ group) has three hydrogens, so there should be four lines with a height of 2. So you should end up with the spectrum shown in Figure 5.12.

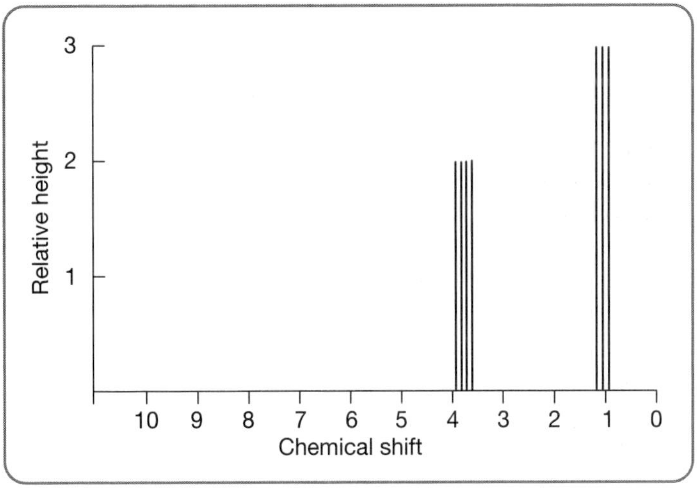

Figure 5.12 Proton NMR spectrum for chloroethane

Example 10

When sodium hydrogen carbonate is heated to 112 °C it decomposes and carbon dioxide gas is given off.

$$2NaHCO_3 \longrightarrow Na_2CO_3 + CO_2 + H_2O$$

The apparatus shown in Figure 5.13 can be used to measure the volume of carbon dioxide produced by the reaction.

a) Why is an oil bath used and not a water bath?

b) Calculate the volume of carbon dioxide produced by the complete decomposition of 0.84 g of sodium hydrogen carbonate.

(Take the molar volume of carbon dioxide to be 24 litre mol^{-1}.)

Show your working clearly.

c) The carbon dioxide could also be collected over water in an inverted measuring cylinder.

Suggest why the above method of collection is likely to be more accurate.

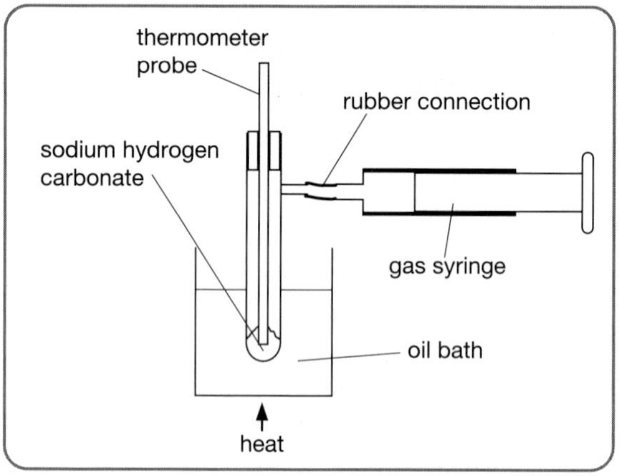

Figure 5.13 Apparatus to measure CO_2 produced

Solution

a) Water boils at 100 °C. It cannot reach a higher temperature. Since the decomposition takes place above 100 °C, a water bath cannot be used. It's not hot enough.

b) 1 mole of $NaHCO_3 = 23 + 1 + 12 + (3 \times 16) = 84$ g

$$2NaHCO_3 \longrightarrow Na_2CO_3 + CO_2 + H_2O$$

2 mole	1 mole
168 g	24 litres
84 g	12 litres
0.84 g	**0.12 litre**

(c) Carbon dioxide is somewhat soluble in water. If collected over water in a measuring cylinder, some would dissolve, and a smaller volume would be collected.

Example 11

A mass spectrometer is an instrument which breaks molecules into fragments and produces a record of the masses of the fragments. From this you can work out the structure of the starting molecule.

Ethane gives the result shown in Figure 5.14

The peak at 15 is produced by a CH_3 fragment. The peak at 30 is produced by the whole, unfragmented molecule.

(a) A compound $C_2H_4Br_2$ can have either of the following structures.

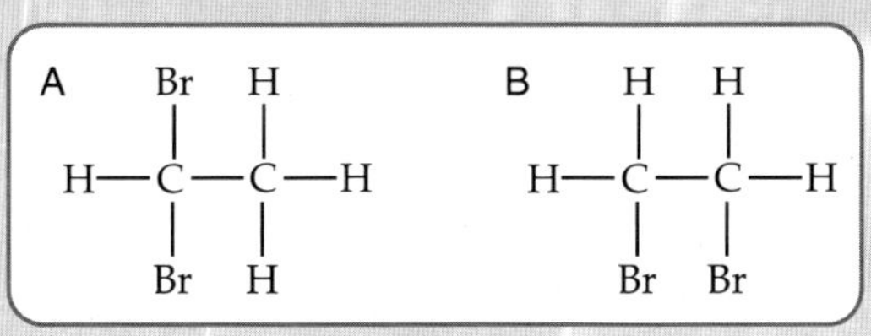

The compound gives peaks at 15, 173 and 188.

Explain which compound, A or B, has been used.

(b) A sample of a halogenated ethane was examined using a mass spectrometer.

The instrument gave a peak at a maximum mass of 48.

Work out the molecular formula for the compound.

Solution

(a) The peak at 15 is due to the group $-CH_3$ $(12 + (3 \times 1))$.

That at 173 is due to $CHBr_2$ and that at 188 is due to the entire, unfragmented molecule. The compound is the **A** isomer.

If it had been B, there would have been peaks at 188 and 94. The peak at 94 would be due to $CBrH_2$. The molecule would split symmetrically.

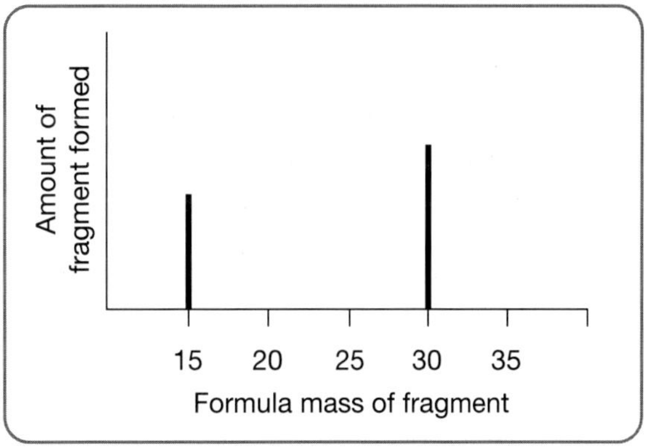

Figure 5.14 Mass spectrum of ethane

(b) If the compound is a halogenated ethane, there are two carbon atoms, with a combined mass of 24. There are only 24 units of mass left to make up the maximum mass of 48. In which case the only halogen atom light enough to be accommodated is fluorine, with a mass of 19. This means also that there must be five hydrogen atoms.

The formula is $\mathbf{CH_3CH_2F}$.

CALCULATIONS

As we said in the Introduction, a big difference between Standard Grade and Higher is that Standard Grade Chemistry contains very few calculations – about 4 or 5 marks out of the 60 marks available in the Credit exam. In the Higher exam, there may be calculations worth up to 25% of the whole exam. So, you have to get lots of practice at chemical calculations so that you're able to tackle them with confidence when it comes to sitting the actual examination. It's normally expected that you'll have Standard Grade Maths at Grade 2 if you are doing Higher Chemistry. If you're weak at Maths then you'll certainly need all the practice you can get!

You'll also need a calculator and you need to know how to use it. It's not a good idea to borrow a calculator on the day of the exam because there's a good chance that it won't be the same as the one you're used to. Check that it's in working order and, if it uses batteries, make sure they're new. Remember, from Standard Grade:

Question: Give a disadvantage of a cell compared with mains electricity.

Answer: A cell can run out!

This is a Standard Grade question you're **not** likely to be asked at Higher but you certainly don't want to be reminded of it in the middle of a calculation.

It's very important to try every calculation in the Higher Chemistry paper, even if you don't know how to complete the calculation. Markers always look for ways of giving you marks – they're not looking for ways of taking marks from you. Lots of calculations are broken down into half mark chunks, and it's usually quite easy to get some of the half marks, even if you don't know how to complete the calculation.

Another thing. Lots of calculations include the instruction:

"**Show your working clearly**"

It's in bold so it looks as if the examiners mean business.

Think about this question.

Starting with a mass of 2·4 g of ethanol, and a slight excess of butanoic acid, a student achieved a 70% yield of the ester ethyl butanoate (mass of one mole = 116 g).

Calculate the mass of ester obtained.

Show your working clearly.

You take out your calculator, do the calculation, and come up with the answer **4·236 g** (which happens to be the right answer!) and that's all you write down on the paper. How many marks will you score out of 2?

Surprise – 2 out of 2!

The reason for this is that a correct answer, without working, gets full marks. You can check this out on the SQA website, where you'll find general marking instructions for markers. Knowing about these instructions is definitely a help in answering questions. This is what the SQA says: *'Full marks are usually awarded for the correct answer to a calculation* ***on its own'*** **–** that is, without working.

Suppose you do the calculation, round off the answer, and write **4·24 g**.

That's OK!

Suppose you do the calculation, round off the answer, and write **4·2 g**.

That's still OK!

Suppose you do the calculation, round off the answer, and write **4 g**.

Not OK! You might not get any marks at all for this – it has the look of a lucky guess, **because there's no working to back it up.**

Also, if you just write 4·236 (or 4·24, or 4·2) you'll lose half a mark. Why? You have to include the correct units in the answer. If the question says 'calculate the mass of ester obtained' you need to give the units of mass. However, if the units are included in the question, like this: 'calculate the mass, in grams, of ester obtained' then you don't have to give the units. But if you do, they need to be correct!

Another important thing – make sure your answer is **reasonable**. Suppose you misread your calculator, or put the decimal point in the wrong place, and come up with the answer 42·36 g (or 423·6 g for that matter). Do you really expect to start with 2·4 g of ethanol and end up with a huge amount of ester? Remember when you made an ester in the lab? How much of the smelly liquid was floating at the top of the mixture in the beaker? Not a lot!

It looked like this. It didn't look like this.

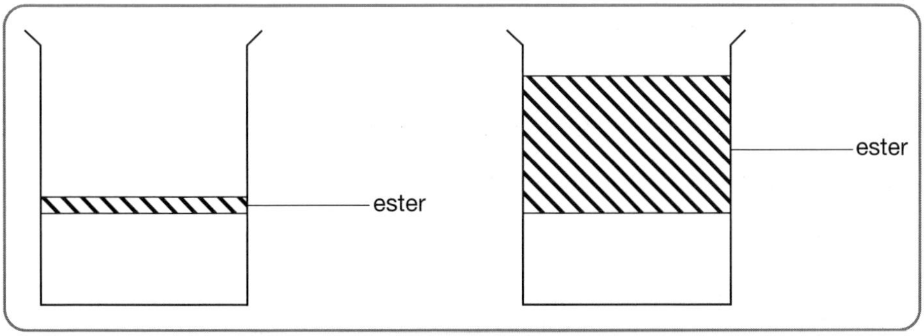

There's another mistake to avoid. The question tells you that one mole of the ester weighs 116 g but you would have needed to work out the mass of one mole of ethanol to do this question. You would have used Relative Atomic Masses from the Data Booklet. Some students might write down as their answer to the ester question **7·495 g**. This is the answer you would get if you made the mistake of using **Atomic Numbers** instead of **Relative Atomic Masses**. This is a serious mistake, and the most you would get for the question would be 1 mark.

Atomic numbers don't turn up in many calculations – usually only in calculations involving radioactivity.

The message in all of this is that it really is important to show your working. Even if you end up with the wrong answer, you get credit for all the correct parts of your calculation. A daft answer can get nearly full marks if most of the steps are right.

Summary

Here's a list of the main kinds of calculations that you're likely to meet.

Unit 1

a) Calculations of reaction rate from numerical data or graphs (usually graphs)

b) Calculations involving enthalpies of combustion, enthalpies of solution or enthalpies of neutralisation

c) Calculations involving the mole and gas volumes

d) Calculation involving excess reactants

Unit 2

a) Percentage yield calculations (often applied to ester formation)

Unit 3

a) Hess's law calculations

b) pH calculations

c) Redox titrations

d) Calculations based on electrolysis

e) Radioactivity calculations

Let's look at examples of each kind of calculation.

Unit 1

Reaction rates

Example 1

The balanced equation for the decomposition of hydrogen peroxide is:

$$2H_2O_2(l) \rightarrow 2H_2O(l) + O_2(g).$$

The graph shown in Figure 6.1 was obtained for the volume of oxygen released over time.

Calculate the average rate of reaction between 20 s and 40 s.

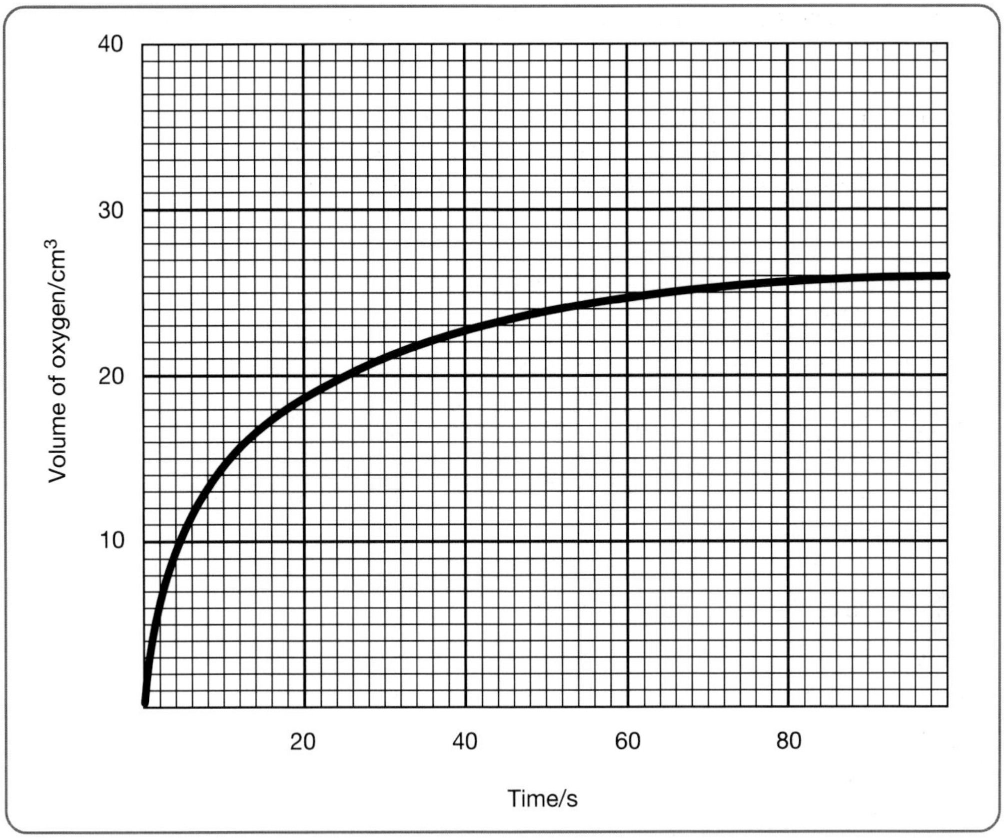

Figure 6.1 Graph of results

Solution

You have to read the graph at 20 seconds. It says 18·5 cm^3.

You have to read the graph at 40 seconds. It says 22·5 cm^3.

The difference is **4 cm^3**.

This difference arose over a period of **20 s**.

The rate is found by dividing the change in volume by the change in time.

So the rate is 4 ÷ 20 or **0·2 $cm^3 s^{-1}$**.

Always make sure you read the graph correctly – be sure you know what each small box is worth. And if you miss out the units – in this case $cm^3 s^{-1}$ – you'll lose half a mark. Remember, units of rate always involve $time^{-1}$, in this case, s^{-1}.

Calculations involving enthalpies of combustion, enthalpies of solution and enthalpies of neutralisation

What You Should Know

It's important to be clear about what these terms mean. (Learn them!)

- The enthalpy of combustion of a substance is the enthalpy change when one mole of the substance burns completely in oxygen.
- The enthalpy of solution of a substance is the enthalpy change when one mole of the substance dissolves in water.
- The enthalpy of neutralisation of an acid is the enthalpy change when the acid is neutralised to form one mole of water.

By far the most common type of enthalpy question is based on enthalpies of combustion, because a question from the Prescribed Practical Activities (PPAs) on enthalpy of combustion can be added on.

Just make sure you know that when substances burn, heat is given out, so ΔH is **negative**! Don't forget that the units used here are **kilojoules per mol** ($\mathbf{kJ\ mol^{-1}}$).

Example 2

When 2·3 g of ethanol (relative formula mass = 46) burns, 68 kJ of energy is released. From this result, calculate the enthalpy of combustion of ethanol.

Solution

The question tells you that one mole of ethanol weighs 46 g. Good – you don't have to work this out! You have been given 2·3 g. To find out how many moles there are, divide the given mass (2·3 g) by the mass of one mole (46 g).

So, 2·3 ÷ 46 = **0·05 mole**.

0·05 mole → 68 kJ

If you divide both numbers by 0·05, it should still be true.

$$\frac{0{\cdot}05}{0{\cdot}05} \rightarrow \frac{68}{0{\cdot}05}$$

So, 1 mole → 1360 kJ.

You mustn't forget that this is heat given out, so the correct answer is:

$$\mathbf{\Delta H = -1360\ kJ\ mol^{-1}}$$

Again, don't forget the sign, and make sure you have included the units.

In another type of question the heat given out by the burning substance is used to heat water. From the temperature rise of the water you can work out the enthalpy of combustion. The formula you have to know is:

$$\textbf{heat} = \mathbf{cm\Delta T}$$

c has the value 4·18. That's the number of kilojoules of heat needed to heat a kilogram of water up by one degree Celsius.

m is the mass of water in kilograms.

$\mathbf{\Delta T}$ is the temperature rise.

Make sure you learn this formula and know what the terms mean.

Example 3

0·23 g of ethanol, C_2H_5OH, was burned and the heat produced used to heat 100 g of water. The temperature of the water rose by 16·3 degrees Celsius.

Calculate the enthalpy of combustion of ethanol.

Solution

The basic thing is to link the moles of ethanol burned to the kJ of heat produced.

So work out the moles of ethanol. To do this, you need to know what one mole of ethanol weighs. You have (2 × 12) + (5 × 1) + 16 + 1 = **46 g**.

There is obviously a lot less than one mole! In fact there is 0·23 ÷ 46 = **0.005** mole.

How much heat was produced? Use:

$$\textbf{heat} = \mathbf{cm\Delta T}$$

This lets you work out that the heat was 4·18 × 0·1 × 16·3 = **6·81 kJ**.

Where does the 0·1 come from? Remember, m is the mass of water in kilograms.

Now you know that **0·005 mole** ⟶ **6·81 kJ**.

It will still be true if you divide both numbers by 0·005.

You divide 0·005 by 0·005, which gives 1·0 mole. You divide 6·81 by 0·005, which gives 1362. This is the amount of heat produced by burning one mole of ethanol. The answer is that the enthalpy of combustion is –**1362 kJ mol**$^{-1}$.

Again, it's important to put down your working, and to do as much of the question as you can. Each item shown above in bold is worth half a mark – you get credit for what you know and what you do!

Example 4

The enthalpy of neutralisation was determined using dilute solutions of hydrochloric acid and sodium hydroxide.

50 cm^3 of 1 mol ℓ^{-1} sodium hydroxide at 20 °C was added to 50 cm^3 of 1 mol ℓ^{-1} hydrochloric acid at 22 °C. After reaction, the temperature was 27·5 °C.

Calculate the enthalpy of neutralisation, in kJ mol^{-1}.

Show your working clearly.

Solution

The heat produced is found from

$$\text{heat} = cm\Delta T$$

where c = 4·18.

m, the mass of liquid heated, is taken to be 100 g (i.e. 0·1 kg). It's assumed that the density of the acid and the alkali are the same as that of water, and that they have the same specific heat capacity of 4·18.

ΔT is the temperature rise. In this example, the starting temperature is taken as the average of 20 and 22, that is, 21 °C. So $\Delta T = 27{\cdot}5 - 21 = 6{\cdot}5$ °C.

$$\text{heat} = 4{\cdot}18 \times 0{\cdot}1 \times 6{\cdot}5 = 2{\cdot}717\,\text{kJ}$$

Remember – the enthalpy of neutralisation of an acid is the enthalpy change when the acid is neutralised to form **one** mole of water.

In this reaction 50 cm^3 of 1 mol ℓ^{-1} solutions were used. These contain 0·05 mole each of H^+ ions and OH^- ions, which form 0·05 mole of water. To obtain the answer scaled up to 1 mole, you divide 2·717 by 0·05 to obtain **54·34 kJ mol^{-1}**.

$$\Delta H = -54{\cdot}34\ \text{kJ mol}^{-1}$$

Calculations involving the mole and gas volumes

There are lots of calculations built round this. However there are just a few basic ideas you need to tackle them and you need to learn these thoroughly. We'll assume that you remember, from Standard Grade, how to calculate the mass of a mole of substance, and how to convert masses to moles or moles to masses. If you don't, then go and revise it **now**.

Summary

These are the basics.

A mole of substance contain 6·02 × 10^{23} units of the substance. The units might be molecules, or a group of ions. For example, you could think of a unit of sodium chloride as consisting of a Na^+ ion and a Cl^- ion (although NaCl, of course, consists of a lattice of these ions). 6·02 × 10^{23} is known as the Avogadro Constant, and it's in the Data Booklet – you don't even have to remember it!

Summary *continued* ➢

Summary continued

Under identical conditions of temperature and pressure, one mole of any gas has the same volume as one mole of any other gas. At room temperature and atmospheric pressure, this is about 24 litres. It follows that the volume of any gas is directly proportional to the number of moles of gas present. Twice the volume, twice the number of moles!

Let's look at one or two examples, to see how to use these ideas.

Figure 6.2 If they're the same size, they hold the same number of moles of gas

Example 5

How many moles of ions are in one mole of magnesium phosphate?

Solution

You need to work out the formula for magnesium phosphate. (No one is going to teach you how to work out formulae in Higher Chemistry. You're expected to be able to do that from Standard Grade.)

It's $Mg_3(PO_4)_2$. How many ions are there? There are three Mg^{2+} ions and there are two PO_4^{3-} ions. That's five ions altogether. So if you have one mole of this stuff, then you have five moles of ions. You should see that this is a simple question. Don't be put off by the introduction of the word 'mole'.

Let's look at a multiple-choice question.

Example 6

The Avogadro Constant is the same as the number of

A atoms in 24 g of carbon

B molecules in 14 g of nitrogen

C molecules in 2 g hydrogen

D ions in one litre of 1 mol ℓ^{-1} potassium chloride solution.

Solution

Remember, the Avogadro Constant represents a mole of substance, so it's just a question about 'How many moles…?'.

A contains two moles of carbon, since the RAM of carbon is 12.

Watch out in B. Nitrogen is diatomic, N_2. This is just half a mole!

C is the right answer. Hydrogen is diatomic, H_2. So one mole of it weighs 2 g.

D won't do. This solution may contain one mole of KCl, but it contains two moles of ions.

A common kind of mole calculation involves working out how much product you can get if you start with a certain amount of reactant.

Example 7

Calculate the mass of carbon dioxide produced when excess hydrochloric acid is added to 5 g of calcium carbonate.

Solution

The equation for this reaction is:

$$CaCO_3 + 2HCl \rightarrow CaCl_2 + H_2O + CO_2$$

No one is going to teach you how to write equations of this kind in Higher – it's Standard Grade work, and you should know it already – if you don't, you should revise the Standard Grade topic on 'Reaction of Acids'.

The first thing to do with the equation is to identify the chemicals which matter.

They're shown below in bold. The others don't matter. Ignore them.

$$\mathbf{CaCO_3} + 2HCl \rightarrow CaCl_2 + H_2O + \mathbf{CO_2}$$

The next thing is to identify the number of moles of chemicals involved.

$$\underset{\text{1 mole}}{\mathbf{CaCO_3}} + 2HCl \rightarrow CaCl_2 + H_2O + \underset{\text{1 mole}}{\mathbf{CO_2}}$$

You know that if you start with one mole of calcium carbonate, you'll end up with one mole of carbon dioxide. Here there is 5 g of calcium carbonate. How many moles is that?

$CaCO_3 = 40 + 12 + (3 \times 16) = 100$. 1 mole weighs 100 g. Here there is 5 g. That is $5 \div 100$ mole = **0.05 mole**.

This means that you will end up with 0·05 mole of carbon dioxide. How much is this?
1 mole of carbon dioxide weighs $12 + (2 \times 16) = 44$ g.

0·05 mole weighs $0{\cdot}05 \times 44 =$ **2.2 g**.

In questions where you're given the mass of a chemical, it's always a good idea to convert this to moles, even if you're unsure how to go further in the question. In a similar way, it's always a good idea to convert a number of moles to a mass. You'll almost certainly get credit for doing this. And sometimes it helps you to see the next step.

Let's look at a selection of gas volume calculations.

Example 8

Under identical conditions of temperature and pressure, which gas occupies the greatest volume?

A 4 g hydrogen

B 4 g helium

C 8 g oxygen

D 64 g sulphur dioxide

Solution

Remember, 1 mole of any gas, under the same conditions of temperature and pressure, has the same volume as one mole of any other gas. This question really just asks: 'Which of these gases contains the most moles?'

It's a good idea in a question like this to write the formula and the formula mass beside each option. Then work out the number of moles given. Be very careful with the formulae – remember – some gases are diatomic (molecules that contain two atoms) and others are monatomic (exist as single atoms). Here goes.

		Formula	Formula mass g	Moles
A	4 g hydrogen	H_2	2	2
B	4 g helium	He	4	1
C	8 g oxygen	O_2	32	0·25
D	64 g sulphur dioxide	SO_2	64	1

Surprise! The answer is **A**, the hydrogen – even though there's only 4 g of it.

You can see that although the new idea about molar volume is quite simple, you have to know quite a lot more to tackle the question successfully.

Example 9

Calculate the volume of oxygen, in litres, required for the complete combustion of one litre of propane gas.

Solution

You won't get far with this if you don't know the formula for propane – it's an alkane, and it fits the general formula C_nH_{2n+2}. But how many carbon atoms does it have?

The useful thing to know here is that the alkanes are listed in the Data Booklet **in increasing order of number of carbon atoms**.

The list in the Data Booklet starts with:

Methane

Ethane

Propane

Propane's third so it has three carbon atoms and its formula is therefore C_3H_8.

The balanced equation is:

C_3H_8	+	$5O_2$	$\rightarrow$	$3CO_2$	+	$4H_2O$
1 mole		5 moles				
1 litre		**5 litres**				

Why? Since one mole of gas has the same volume as one mole of any other gas, it follows that the volumes are directly proportional to the moles.

Calculations involving excess reactants

Summary

- These questions are usually tackled by comparing the numbers of moles supplied with the mole ratio in the equation.
- It almost always involves converting masses to moles.
- It sometimes involves converting volumes and concentrations to moles.

Example 10

A chemist added 4 g of magnesium powder to 100 cm^3 of 2·0 mol ℓ^{-1} hydrochloric acid. The balanced equation for this reaction is:

$$Mg + 2HCl \rightarrow MgCl_2 + H_2$$

Show by calculation which reactant is in excess.

Show your working clearly.

Solution

The equation tells you that one mole of magnesium reacts with two moles of acid.

You need to convert the given quantities of these chemicals to moles and compare the numbers of moles.

1 mole of magnesium weighs about 24 g.

4 g is therefore 4/24 mole = **0·17 mole**.

100 cm^3 is 0·1 litre. The moles of acid = volume (litres) × concentration (mol ℓ^{-1})

This is 0·1 × 2 = **0·2 mole**.

There are more moles of the acid. Does this mean that it's in excess? Look at the equation – you need **two** moles of acid for every mole of magnesium.

You actually need **2 × 0·17 = 0·34 mole** of the acid, more than you have been given. There's not enough acid – it's the magnesium which is in excess!

You can have similar questions using gas volumes.

Example 11

One litre of propane was burned in the presence of four litres of oxygen. What gases would be present at the end of the reaction if the temperature is 150 °C?

Solution

The equation for this reaction is:

C_3H_8	+	$5O_2$	$\rightarrow 3CO_2 + 4H_2O$
1 mole		5 moles	
1 litre		**5 litres**	

If you want to burn one litre of propane, you need five litres of oxygen. You have only been given four – not enough to burn all the propane, so some will be left at the end.

Therefore, the gases present at the end of the reaction are **propane**, **carbon dioxide** and **water** (remember, above 100 °C, water is a gas).

Example 12

Butane reacts with an oxide of nitrogen.

$$C_4H_{10}(g) + 13N_2O(g) \rightarrow 4CO_2(g) + 5H_2O(g) + 13N_2(g)$$

a) What volume of nitrogen will form from the complete reaction of 10 cm^3 of butane?

b) The mixture of product gases is passed through a solution of sodium hydroxide at 20 °C prior to collection. Only nitrogen gas is collected.

   Explain why

   i) $H_2O(g)$ is not collected.

   ii) $CO_2(g)$ is not collected.

Solution

a) Remember, the volumes of gases are proportional to the number of moles present. So you'll get 130 cm^3 of nitrogen.

b) (i) Because the solution of sodium hydroxide is at 20 °C, any water vapour formed in the reaction would condense to water.

   ii) Sodium hydroxide is an alkali. Since carbon dioxide is the oxide of a non-metal, it is acidic. The carbon dioxide, therefore, will react with the sodium hydroxide and cannot be collected.

Example 13

The apparatus shown in Figure 6.3 was used to study the reaction of hydrocarbon gases with oxygen at room temperature.

A high voltage is used to generate a spark between the platinum electrodes.

In one experiment, 20 cm^3 of a hydrocarbon gas containing four carbon atoms per molecule was ignited in excess oxygen gas. Carbon dioxide and water vapour were produced.

a) Calculate the volume of carbon dioxide produced.

b) 60 cm^3 of water vapour was produced.

What is the molecular formula for the hydrocarbon?

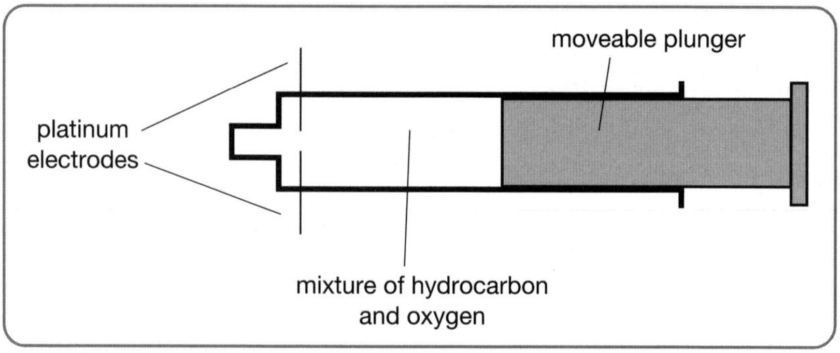

Figure 6.3 Apparatus to study reaction of hydrocarbon gases with oxygen

Solution

a) Each mole of carbon atoms will produce one mole of carbon dioxide.

This means that one mole of gas, when burned, will produce four moles of carbon dioxide. Since the volumes of gases are proportional to the number of moles of gas, 20 cm^3 of hydrocarbon will produce 80 cm^3 of carbon dioxide.

b) Using the reasoning above, three moles of water must form from one mole of hydrocarbon. This means that one mole of the hydrocarbon must contain six hydrogen atoms since each water molecule contains two.

The formula is $\mathbf{C_4H_6}$.

Unit 2

Percentage yield calculations

Summary

Often a percentage yield calculation involves ester formation or a similar type of reaction, since such reactions are reversible and only go to equilibrium. As a result, there isn't 100% conversion of reactants to products.

These are really just simple calculations based on the amount of product expected from a given amount of reactant. If the amount of product is less than predicted, then you can calculate the percentage of the expected amount. That's the percentage yield.

Example 14

Paracetamol, the pain killer, has the following structure.

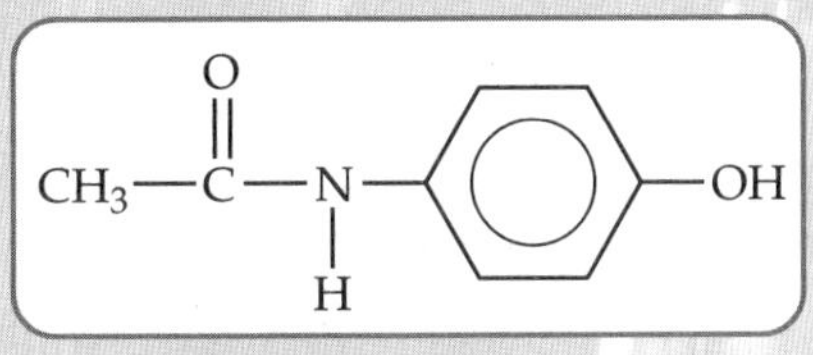

It can be formed from the following two compounds.

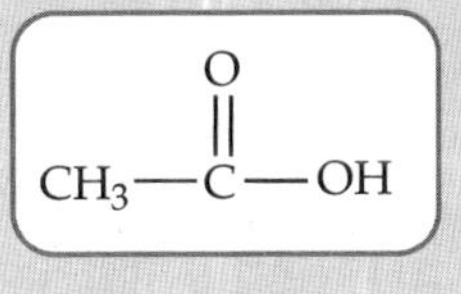

Compound A

and

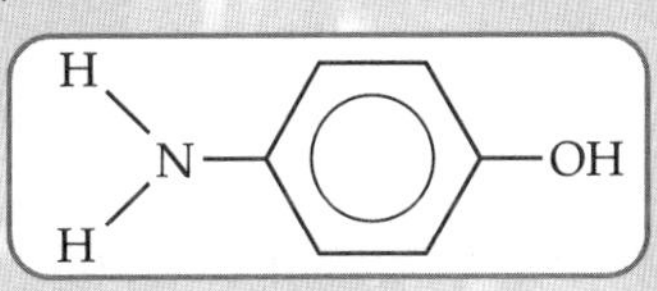

Compound B

5·4 g of compound B was made to react with an excess of compound A. 6·04 g of paracetamol formed.

Calculate the percentage yield of paracetamol.

Show your working clearly.

Solution

First, you have to be clear about what's going on here. The reaction is simply:

$$\text{A} + \text{B} \longrightarrow \text{paracetamol}$$

$$\text{1 mole} \quad \text{1 mole} \qquad\quad \text{1 mole}$$

There's an excess of A, so you can forget about it. It won't affect how much product there is.

What does one mole of B weigh? The benzene ring has six carbon atoms and, in this compound, four hydrogen atoms. One mole weighs $(6 \times 12) + 4 + 17 + 16$. The 17 is from the –OH group and the 16 is from the $-NH_2$ group. The total is **109 g**.

In a similar way, you should find that one mole of paracetamol weighs **151 g**. Check it.

If you've done this correctly, you will have found that:

$$109\,g \text{ of B} \rightarrow 151\,g \text{ paracetamol}$$

If you divide both numbers by 109, it will still be true. This will give:

$$1\,g \text{ of B} \rightarrow 151 \div 109 = 1{\cdot}385\,g \text{ paracetamol}$$

If you multiply both numbers by 5·4, it will still be true.

$$5{\cdot}4\,g \text{ of B} \rightarrow 1{\cdot}385 \times 5{\cdot}4 = \mathbf{7{\cdot}48\,g} \text{ paracetamol}$$

7·48 g of paracetamol is the amount of product you'd expect. This is what's called the **theoretical yield**. The question asks for the **actual yield**. This was 6·04 g.

The percentage yield is **actual yield** × **100** ÷ **theoretical yield** = **6·04** × **100** ÷ **7·48** = **80·7%**.

Unit 3

Hess's law calculations

Summary

Hess's law lets you use enthalpy changes for given reactions to work out enthalpy changes for other reactions.

Look at this equation.

$$C + O_2 \rightarrow CO_2 \qquad \Delta H = -394\,kJ\,mol^{-1}$$

It tells you that if you burn one mole of carbon completely in oxygen, 394 kJ of heat are released. Obviously, if you burn two moles you would expect twice as much heat to be given out. If you start with the product, CO_2, and break it down into carbon and oxygen, you'd expect to put in 394 kJ to do this. **Anything you do to the equation, you must do to the value of ΔH.**

If you add equations, to get new ones, you must add the ΔH values.

If you multiply equations, you must multiply the ΔHs.

If you reverse an equation, you must reverse the ΔH – that is, change its sign.

Example 15

Here is information on two combustion reactions.

Equation 1 $CO + \frac{1}{2}O_2 \rightarrow CO_2 \qquad \Delta H = -283\,kJ\,mol^{-1}$

Equation 2 $Cu + \frac{1}{2}O_2 \rightarrow CuO \qquad \Delta H = -155\,kJ\,mol^{-1}$

Use this information to find the enthalpy change for the reaction:

$$CuO + CO \rightarrow Cu + CO_2$$

Solution
You can call the final above equation the 'target equation'.

The key to juggling equations 1 and 2 is to put the chemicals on the side of their equations that matches the positions of the chemicals in the target.

Look at equation 1. Does the CO match the position of the CO in the target? Yes, it does – they're both on the left-hand side.

Look at equation 2. Does the Cu match the position of the Cu in the target? No, it doesn't, so you'll have to reverse the equation and change the sign of ΔH.

Now you have:

$$CO + \frac{1}{2}O_2 \rightarrow CO_2 \qquad \Delta H = -283\ kJ\ mol^{-1}$$

$$CuO \rightarrow Cu + \frac{1}{2}O_2 \qquad \Delta H = +155\ kJ\ mol^{-1}$$

Now you've done as much as you can to make the equations match the target. We can put them together and hope we've hit the target! You can cancel out chemicals which appear in the **same** quantities which appear on **both** sides of the arrows. Like this:

$$CO + \cancel{\tfrac{1}{2}O_2} \rightarrow CO_2 \qquad \Delta H = -283\ kJ\ mol^{-1}$$

$$CuO \rightarrow Cu + \cancel{\tfrac{1}{2}O_2} \qquad \Delta H = +155\ kJ\ mol^{-1}$$

$$CuO + CO \rightarrow Cu + CO_2 \qquad \Delta H = \mathbf{-128\ kJ\ mol^{-1}}.$$

Here's a slightly harder example.

Example 16

Given the following equations and enthalpies

Equation 1 $\quad H_2 + \frac{1}{2}O_2 \rightarrow H_2O \qquad \Delta H = -286\ kJ\ mol^{-1}$

Equation 2 $\quad C + O_2 \rightarrow CO_2 \qquad \Delta H = -394\ kJ\ mol^{-1}$

Equation 3 $\quad C_2H_6 + 3\frac{1}{2}O_2 \rightarrow 2CO_2 + 3H_2O \qquad \Delta H = -1560\ kJ\ mol^{-1}$

work out the enthalpy change for this reaction.

$2C + 3H_2 \rightarrow C_2H_6$ (this is the target equation)

Solution
Again, it's a case of matching the equations to the target. You need three times as much hydrogen as shown in Equation 1, so you must multiply Equation 1 by three. You must also multiply ΔH by three.

You should see that you must multiply Equation 2 by two and also reverse Equation 3.

This is what you end up with:

$$3H_2 + 1\tfrac{1}{2}O_2 \rightarrow 3H_2O \qquad \Delta H = -858\ \text{kJ mol}^{-1}$$

$$2C + 2O_2 \rightarrow 2CO_2 \qquad \Delta H = -788\ \text{kJ mol}^{-1}$$

$$2CO_2 + 3H_2O \rightarrow C_2H_6 + 3\tfrac{1}{2}O_2 \qquad \Delta H = +1560\ \text{kJ mol}^{-1}$$

Then you can cancel, as before, and add.

$$3H_2 + \cancel{1\tfrac{1}{2}O_2} \rightarrow \cancel{3H_2O} \qquad \Delta H = -858\ \text{kJ mol}^{-1}$$

$$2C + \cancel{2O_2} \rightarrow \cancel{2CO_2} \qquad \Delta H = -788\ \text{kJ mol}^{-1}$$

$$\cancel{2CO_2} + \cancel{3H_2O} \rightarrow C_2H_6 + \cancel{3\tfrac{1}{2}O_2} \qquad \Delta H = +1560\ \text{kJ mol}^{-1}$$

$$2C + 3H_2 \rightarrow C_2H_6 \qquad \mathbf{\Delta H = -86\ kJ\ mol^{-1}}$$

You have reached the target equation.

It's really important to set down your working in a question like this.

If you haven't got the right answer of –86 kJ mol^{-1}, markers will be looking for the numbers –858, –788 and +1560 on your script, as evidence that you've gone correctly through some of the stages. It's not even necessary to write down all the equations (unless you're asked to), but you must show evidence that you've processed at least some of the data.

Example 17

What is the relationship between enthalpies p, q, r and s?

$$S + H_2 \rightarrow H_2S \qquad \Delta H = p$$

$$H_2 + \tfrac{1}{2}O_2 \rightarrow H_2O \qquad \Delta H = q$$

$$S + O_2 \rightarrow SO_2 \qquad \Delta H = r$$

$$H_2S + 1\tfrac{1}{2}O_2 \rightarrow H_2O + SO_2 \qquad \Delta H = s$$

A $p = q + r - s$

B $p = s - q - r$

C $p = q - r - s$

D $p = s + r - q$

Solution

This question is asking you to rearrange the last three equations in order to arrive at the first equation. The enthalpy for the first equation is found by treating the other three enthalpies in the same way as the equations.

You can see that in the first equation, for which $\Delta H = p$, sulphur and hydrogen appear on the left-hand side of the equation. That means that the second and third equations have these reactants in the correct position (on the left).

In the last equation, the hydrogen sulphide is on the left. However, in the first equation, it is on the right. So the last equation must be reversed.

Don't worry about the oxygens – they'll take care of themselves by cancelling out.

Since the second and third equations do not need to be changed, q and r must both be positive. Look at answer A. Is that the one? You had to reverse the last equation. Therefore s must be negative. Answer **A** is certainly the correct choice!

pH calculations

Summary

You don't have much to learn for pH calculations. Just remember that the concentration of H^+ ions multiplied by the concentration of OH^- ions is always 1×10^{-14}. This is sometimes written like this: $[H^+][OH^-] = 1 \times 10^{-14}$. The square brackets are shorthand for 'concentration in moles per litre'.

You also have to remember the link between pH values and the concentration of H^+ ions. You probably have filled in a table that looks like this.

$[H^+]$/mol ℓ^{-1}	pH
1×10^{0}	0
1×10^{-1}	1
1×10^{-2}	2
1×10^{-3}	3
⋮	⋮
1×10^{-14}	14

Also, make sure that you realise that 0·1 can be written 1×10^{-1}, 0·01 can be written 1×10^{-2} etc.

Most of the questions you meet on pH are likely to be quite easy.

Example 18

A solution has a pH of 6.

What is the concentration of H^+ ions in this solution?

Solution
1×10^{-6} mol ℓ^{-1}

The question could have asked 'What is the concentration of OH^- ions in this solution?' Then you would have had to remember that $[H^+][OH^-] = 1 \times 10^{-14}$, so it would be necessary to divide 1×10^{-14} by 1×10^{-6} (the concentration of H^+ ions), obtaining **1×10^{-8}**, the concentration of the OH^- ions, in mol ℓ^{-1}.

Redox titration calculations

Summary

These mostly involve working with ion–electron equations, or occasionally with a full, balanced redox equation.

Example 19

The ion–electron equations for the redox reaction between iodide ions and dichromate ions are:

$$2I^- \rightarrow I_2 + 2e^-$$

$$Cr_2O_7^{2-} + 14H^+ + 6e^- \rightarrow 2Cr^{3+} + 7H_2O$$

How many moles of iodide are oxidised by one mole of dichromate ions? (This might appear as a multiple-choice question with four options.)

Solution

The top equation tells you that two moles of iodide ions lose two moles of electrons. (This is oxidation). This means that one mole of iodide ions lose one mole of electrons.

The dichromate ion in the second equation can accept six moles of electrons. This means that one mole of dichromate ions is capable of oxidising six moles of iodide ions. The answer is **6**.

Example 20

The ion–electron equations for the redox reaction between sulphite ions and permanganate ions are:

$$SO_3^{2-} + H_2O \rightarrow SO_4^{2-} + 2H^+ + 2e^-$$

$$MnO_4^- + 8H^+ + 5e^- \rightarrow Mn^{2+} + 4H_2O$$

How many moles of sulphite are oxidised by one mole of permanganate ions?

Solution

The top equation tells you that sulphite ions can lose two moles of electrons.

The bottom equation tells you that the permanganate ions can accept five moles of electrons. This means that one mole of permanganate ions is capable of oxidising 2·5 moles of sulphite ions. The answer is **2·5**.

You could also be asked to use data from a redox titration to work out the concentration of one of the reactants. In this case, you'll almost certainly be given the complete balanced equation for the reaction.

Example 21

An acidified solution of permanganate ions can be used to oxidise sulphite ions to sulphate ions.

The equation for the reaction is:

$$2MnO_4^- + 5SO_3^{2-} + 6H^+ \rightarrow 2Mn^{2+} + 5SO_4^{2-} + 3H_2O$$

In one experiment, 20 cm^3 of a solution containing sulphite ions required 10 cm^3 of 0·02 mol ℓ^{-1} potassium permanganate for complete reaction.

Calculate the concentration of the sulphite ions.

Show your working clearly.

Solution

There's more than one way of doing this kind of calculation.

Method 1

Work out the number of moles of permanganate used.

Moles = volume (litres) × concentration (mol ℓ^{-1}) = 0·01 × 0·02 = 0·0002.

The equation tells you that two moles of permanganate react with five moles of sulphite.

To get the moles of sulphite, you multiply the moles of permanganate by $2\frac{1}{2}$.

The moles of sulphite are $2\frac{1}{2}$ × 0·0002 = 0·0005.

0·0005 mole of sulphite is contained in 0·02 litre. So the concentration in mol ℓ^{-1} is 0·0005 ÷ 0·02 = **0·025 mol ℓ^{-1}**.

Method 2

You can call this method the 'mole ratio' method.

The mole ratio between permanganate and sulphite is 2 to 5.

$$2MnO_4^- + 5SO_3^{2-} + 6H^+ \rightarrow 2Mn^{2+} + 5SO_4^{2-} + 3H_2O$$

2 moles 5 moles

The number of moles is found by multiplying the volume, V, (litres) by the concentration, M, (moles per litre). We'll use the subscript 'p' to stand for permanganate and the subscript 's' to stand for sulphite.

Now you can write:

$$2MnO_4^- + 5SO_3^{2-} + 6H^+ \rightarrow 2Mn^{2+} + 5SO_4^{2-} + 3H_2O$$

2 moles 5 moles

V_pM_p V_sM_s

Then you can say that $\frac{V_pM_p}{V_sM_s} = \frac{2}{5}$.

You can then cross-multiply and get $5V_pM_p = 2V_sM_s$.

We can rearrange this to give $M_s = \frac{5V_pM_p}{2V_s}$ = $\mathbf{0{\cdot}025\ mol\ \ell^{-1}}$.

Method 3

You can call this method the 'PVC' method!

To use this, you need to know the ion–electron equations for the oxidation and the reduction processes involved in the reaction. To do this, you look at the Electrochemical Series in the Data Booklet, identify the equations, and note the number of electrons involved in the relevant ion–electron equations.

You'll find that the sulphite equation involves two electrons and the permanganate equation involves five electrons. We say the 'power' of sulphite is 2 and the 'power' of permanganate is 5.

You can write $P_pV_pC_p = P_sV_sC_s$.

(C is the concentration in mol ℓ^{-1} – it's the same as M above, and you could learn it as the PVM method if you like!)

Then $5 \times 0{\cdot}01 \times 0{\cdot}02 = 2 \times 0{\cdot}02 \times C_s$.

So $0{\cdot}001 = 0{\cdot}04C_s$ giving $C_s = 0{\cdot}001 \div 0{\cdot}04 =$ $\mathbf{0{\cdot}025\ mol\ \ell^{-1}}$.

Calculations similar to these can be used to find the mass of a substance as well as its concentration.

Example 22

Vitamin C is a reducing agent. It has the formula $C_6H_8O_6$. The concentration of a solution of vitamin C can be found by titration with a standard solution of iodine. The indicator used is starch.

The equation is:

$$C_6H_8O_6 + I_2 \rightarrow C_6H_6O_6 + 2H^+ + 2I^-$$

In an experiment, a tablet of vitamin C was dissolved in some water and the solution was made up to 250 cm^3. 25 cm^3 portions of this solution were found to react completely with an average of 28·5 cm^3 of 0·021 mol ℓ^{-1} iodine solution.

Calculate the mass of vitamin C present in the tablet.

Solution

If you're told a volume of a solution and its concentration and you're not sure what else to do, multiply them to find the number of moles. Remember, the volume must be changed to litres!

Number of moles of iodine = $0{\cdot}0285 \times 0{\cdot}021 = 0{\cdot}000\,5985$

which can be written as $5{\cdot}99 \times 10^{-4}$.

Now you have to look at the equation to find the mole ratio in which iodine and vitamin C react. It's nice and simple – one to one. So the moles of vitamin C in the sample are also $5{\cdot}99 \times 10^{-4}$.

There's a slight catch in this question. This number of moles refers to a 25 cm^3 sample. However, the whole tablet was dissolved and made up to 250 cm^3. Therefore, you will have to multiply this number of moles by 10.

The total number of moles of vitamin C is $5{\cdot}99 \times 10^{-3}$. From this you can find the mass, but first you'll have to work out the mass of one mole of vitamin C.

The formula is $C_6H_8O_6$; one mole weighs $(6 \times 12) + 8 + (6 \times 16) = 176$.

The mass of the tablet, therefore, is $5{\cdot}99 \times 10^{-3} \times 176 =$ **1·054 g**.

This is a reasonable answer, because vitamin C tablets are often around 1·0 g in weight.

Remember, always try to decide if your answer is reasonable – it certainly wouldn't be reasonable for the tablet to weigh about 10 g (or 0·01 g for that matter!).

Electrolysis calculations

Summary

Electrolysis calculations nearly always involve the same set of routines. As you know, when you electrolyse a substance, you use electricity to generate products at the positive and negative electrodes. Just how much product you get depends on how much electricity you've sent through the wires.

The important thing to learn is that the amount of electricity, the 'charge' is measured in units called 'coulombs'.

coulombs = current (amps) × time (seconds)

Then, if you divide the coulombs by the number 96 500 you find the number of moles of electrons which have passed through the substance. The moles of electrons are sometimes referred to as 'faradays'. Then if you know the ion–electron equation for the reaction you're interested in, you can work out how many moles of product have formed. No matter what units the time is measured in, you have to change it to seconds.

Even if you are not sure how to complete the calculation, **always** calculate the number of coulombs, and go on to calculate the number of faradays.

Example 23

Calculate the mass of copper formed when a current of 0·5 amps is passed through a solution of copper sulphate for 30 minutes.

Solution

First, you need to calculate the number of coulombs.

Charge = amps × seconds = 0·5 × 30 × 60 = 900 coulombs.

Moles of electrons = 900 ÷ 96 500 = 0·0093.

Now you need the ion–electron equation for the reaction which forms the copper. You'll either be told this, or you'll find it in the Data Booklet.

$$Cu^{2+} + 2e^- \rightarrow Cu$$

This equation tells you two moles of electrons are needed to form one mole of copper. The moles of copper are half the number of moles of electrons. So the moles of copper formed are 0·0093 ÷ 2 = 0·0047.

1 mole of copper weighs 63·5 g.

0·0047 mole of copper weighs 63·5 × 0·0047 = **0·296 g**.

Example 24

A student electrolysed dilute acid in order to calculate the molar volume of hydrogen gas. A current of 0.5 amps was passed for 15 minutes and the volume of hydrogen collected was 55 cm^3.

Solution

Calculate the number of coulombs.

Charge = amps × seconds = 0·5 × 15 × 60 = 450 coulombs.

Moles of electrons = 450 ÷ 96 500 = 0·0047.

The ion–electron equation for the production of hydrogen is:

$$2H^+ + 2e^- \rightarrow H_2$$

It takes two moles of electrons to make one mole of hydrogen.

Therefore the number of moles of hydrogen formed is 0·0023.

This question is a bit different from the previous one, because you have now to work out the molar volume of hydrogen. This just means that you have to work out the volume occupied by a mole of the gas.

0·0023 moles occupy 55 cm^3.

The relationship will still be true if you divide both numbers by 0·0023.

This gives:

1 mole occupies 23 589 cm^3 or **23·59 litre**.

If you know that, at room temperature and pressure, the molar volume of a gas is about 24 litres, then you're able to tell that this is a reasonable answer. Be suspicious of any answer to a calculation of molar volume where the answer is miles away from 24 litres.

Example 25

In the production of copper by electrolysis of a solution of copper ions, the reaction taking place is:

$$Cu^{2+} + 2e^- \longrightarrow Cu$$

How much electricity is required to form 0·4 mole of copper?

A 19 300 C

B 38 600 C

C 77 200 C

D 154 400 C

Solution

The question tells you that 0·4 mole of copper is to be made. From the equation, you can see that to make one mole of copper, you need two moles of electrons. Therefore to make 0·4 moles of copper, you need 0·8 mole of electrons.

You're expected to know that one mole of electrons is 96 000 C – the Faraday constant, which is given in the Data Booklet. All you have to do is multiply this number by 0·8 and you've got the answer.

$0.8 \times 96\,500 = 77\,200$. The answer is **C**.

Sometimes, an electrolysis question is about an industrial process, which involves much bigger numbers. You need to take special care with these questions, because there's more chance of making errors.

Example 26

Zinc can be produced electrolytically from molten zinc salts.

The ion–electron equation for the process is:

$$Zn^{2+} + 2e^- \longrightarrow Zn$$

Calculate the mass of zinc formed if a current of 2000 A is passed for 24 hours.

Solution

The routine is the same.

First, you calculate the number of coulombs.

Charge = amps × seconds = $2000 \times 24 \times 60 \times 60 = 1{\cdot}728 \times 10^8$.

Moles of electrons = $1{\cdot}728 \times 10^8 \div 96\,500 = 1{\cdot}791 \times 10^3$.

The equation tells you two moles of electrons are needed to form one mole of zinc. The moles of zinc are half the number of moles of electrons. So the moles of zinc formed are $1{\cdot}791 \times 10^3 \div 2 = 8{\cdot}95 \times 10^2$.

1 mole of zinc weighs 65·4 g.

$8{\cdot}95 \times 10^2$ mole of zinc weighs $65{\cdot}4 \times 8{\cdot}95 \times 10^2 = \mathbf{5{\cdot}86 \times 10^4\,g}$.

Example 27

If 96 500 C of electricity are passed through separate solutions of copper(II) chloride and nickel(II) chloride, then:

A equal masses of copper and nickel will form

B the same number of atoms of each metal will be deposited

C the metals will be plated on the positive electrode

D different numbers of moles of each metal will be deposited.

Solution

You can forget C. Your Standard Grade work should tell you that the metals will be deposited on the negative electrode – after all, metal ions have a positive charge.

Since copper and nickel have different relative atomic masses, it's very unlikely that equal masses will be deposited. That leaves B and D. If you think carefully about this, they are 'mutually exclusive' – that is, each rules the other out because the number of atoms of a metal is directly proportional to the number of moles. One of these has to be the right answer. Since you are passing the same number of coulombs through the solutions, you are passing the same number of moles of electrons through the solutions. Because both metals form 2+ ions, you expect to get the same number of moles of the metals, and hence the same number of atoms. **B** is the answer.

Radioactivity calculations

Summary

There aren't many different kinds of calculation based on radioactivity. Writing balanced nuclear equations involves some simple arithmetic.

Example 28

Write a balanced nuclear equation for the alpha decay of plutonium-242.

Solution

Even if you can't remember what is meant by alpha decay, don't worry! You can find it in the Data Booklet. Look at the page of radioactive decay series.

RADIOACTIVE DECAY SERIES

Note In both tables γ emissions have been omitted.

TABLE 1 (Plutonium-Uranium)

Element	Symbol	Mass Number	Atomic Number	Type of Radiation	Half-life Period
plutonium	Pu	242	94	α	$3{\cdot}79 \times 10^5$ years
uranium	U	238	92	α	$4{\cdot}51 \times 10^9$ years
thorium	Th	234	90	β	24·1 days
protactinium	Pa	234	91	β	6·75 hours
uranium	U	234	92	α	$2{\cdot}47 \times 10^5$ years
thorium	Th	230	90	α	$8{\cdot}0 \times 10^4$ years

Figure 6.4 Part of the radioactive decay series from the Data Booklet

You can see that the atomic number falls by 2 and the mass number falls by 4. So provided you know that the mass number is written top left and the atomic number is written bottom left, you should be able to do this.

You get $^{242}_{94}Pu \rightarrow ^{238}_{92}U + ^{4}_{2}He$.

Another kind of calculation you can meet involves half-life. You should know that the half-life is the time it takes for the radioactivity to fall by 50%. Incidentally, you should also know that the half-life is a constant for a particular isotope – it cannot be changed – so if you meet a question which asks you how some factor or other affects the half-life – **it doesn't**!

Example 29

The half-life of tritium, a radioactive isotope of hydrogen, is 12·3 years.

How long will it take for the radioactivity of a sample of tritium to fall to less than 5% of its original level?

Solution

This is just a case of finding how many half-lives will take the radioactivity below this value. You could make a table (or count the half-lives on your fingers!)

1 half-life – 50%

2 half-lives – 25%

3 half-lives – 12.5%

4 half-lives – 6.25%

So five half-lives will definitely get you below 5%.

The time required will be 5 × 12·3 years = **61·5 years**.

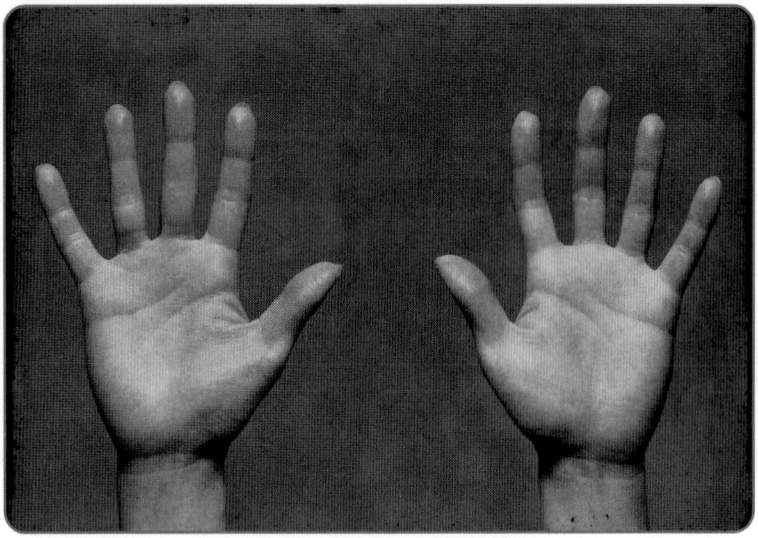

Figure 6.5 Enough fingers for a half-life calculation

Chapter 7

PRESCRIBED PRACTICAL ACTIVITIES (PPAs)

Section B of your exam contains six marks of questions on PPAs, two marks from each unit. Easy to get, if you learn your stuff! The exam tests purely practical aspects of the PPAs – not theoretical detail. You must know the practical details thoroughly – what chemicals are used, what was done, what measurements or observations are made, and safety precautions.

Here's a summary of the PPAs.

Unit 1

Effect of concentration on reaction rate

This experiment studies the reaction of hydrogen peroxide with iodide ions.

The reaction is:

$$H_2O_2 + 2H^+ + 2I^- \rightarrow 2H_2O + I_2$$

The mixture also contains starch and sodium thiosulphate. The sodium thiosulphate reacts with the iodine.

$$I_2 + 2S_2O_3^{2-} \rightarrow 2I^- + S_4O_6^{2-}$$

When all the thiosulphate has reacted, iodine accumulates and the starch turns blue/black. The time taken for the colour change is noted.

You use the same volumes of all chemicals, except potassium iodide. You use decreasing volumes of potassium iodide, replacing it with water so that the concentrations of the other reactants stay the same.

The rate is calculated as 1/time.

The sharp colour change at the end, and the time for the change, allows accurate measurement.

Effect of temperature on reaction rate

This involves the reaction between oxalic acid and acidified potassium permanganate.

$$5(COOH)_2 + 6H^+ + 2MnO_4^- \rightarrow 2Mn^{2+} + 10CO_2 + 8H_2O$$

Oxalic acid solution is added to a mixture of potassium permanganate and sulphuric acid. The permanganate changes from purple to colourless.

All the experiments are identical except for the temperature.

The mixture of sulphuric acid, potassium permanganate and water is heated to the required temperature. The beaker is placed on a white tile. Oxalic acid is added, and the mixture stirred with a thermometer. The reaction time is measured. The temperature is re-measured and an average taken.

The rate is calculated as 1/time.

The experiment is repeated at higher temperatures.

Note:

- The experiments are carried out in dry beakers to avoid affecting the concentrations.
- An accurate experiment is not possible at room temperature, because the gradual colour change makes it difficult to judge when the mixture becomes colourless.

Enthalpy of combustion of ethanol

You use the apparatus shown in Figure 7.1.

The burner is weighed.

A known mass of water is placed in a copper can and its temperature measured.

The burner is placed below the can and ignited.

Once the temperature rises by a certain amount, the burner is extinguished and reweighed.

You collect the following data:

- initial and final mass of burner (gives mass of alcohol burned)
- initial and final temperature of the water (gives temperature rise, ΔT)
- mass of water, assuming 1 g water = 1 cm^3.

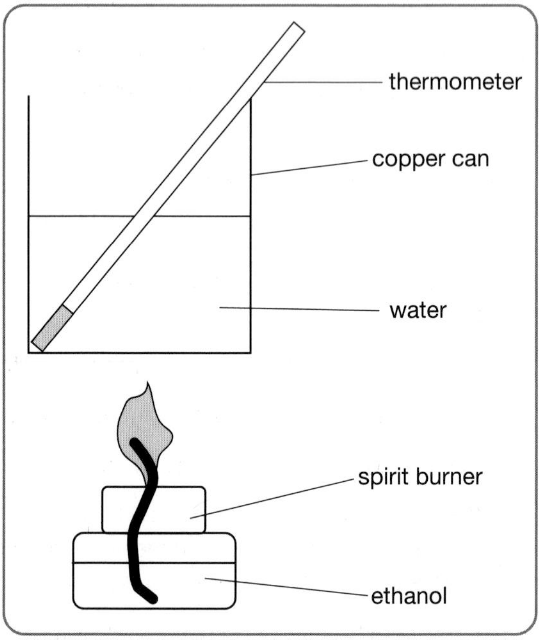

Figure 7.1 Apparatus for determining the enthalpy of combustion of ethanol

Calculation

The heat transferred = $cm\Delta T$ which is scaled up to 1 mole of ethanol.

The main sources of error are:

- incomplete combustion of the ethanol
- heat loss to surroundings
- evaporation of ethanol from the wick.

If you're asked to draw the apparatus, include what's shown in Figure 7.1, plus a draught shield.

Unit 2

Oxidation

This PPA is about tests which distinguish aldehydes from ketones.

Aldehydes are easier to oxidise than ketones. The following oxidising agents are used to distinguish them:

- Benedict's solution (blue, due to Cu^{2+} ions)
- Tollens' reagent (contains ammonia and silver ions)
- acidified potassium dichromate solution.

The compounds and reagents are heated in test tubes in a hot water bath because they are flammable.

Aldehydes cause the following changes:

- Benedict's solution becomes **orange/red**.
- Tollens' reagent gives a **silver mirror** on the wall of the test tube.
- Acidified potassium dichromate solution changes **from orange to green/blue**.

Ketones do **not** affect these reagents.

Esters

Alcohol and carboxylic acid are placed in a test tube with concentrated sulphuric acid catalyst.

A wet paper towel is used as a condenser at the top, preventing loss of volatile ester.

The flammable mixture must be heated with a water bath.

A plug of cotton wool prevents liquid spurting out of the tube.

After a time, the mixture is poured into a beaker of sodium hydrogen carbonate solution. This neutralises the catalyst. Unreacted acid and alcohol dissolve and the ester floats as a separate layer on the surface. It has a sweet, fruity odour.

Factors affecting enzyme activity

Catalase, contained in potato discs, catalyses the breakdown of hydrogen peroxide to water and oxygen.

$$2H_2O_2 \longrightarrow 2H_2O + O_2$$

The apparatus shown in Figure 7.2 is used.

After three minutes hydrogen peroxide solution is added from a syringe.

The number of bubbles of gas produced in one minute is noted.

The experiment is repeated using solutions of other pHs, and also 0.1 mol ℓ^{-1} HCl (pH = 1) and 0.1 mol ℓ^{-1} NaOH (pH = 13).

This experiment can also be done, varying the temperature instead of the pH.

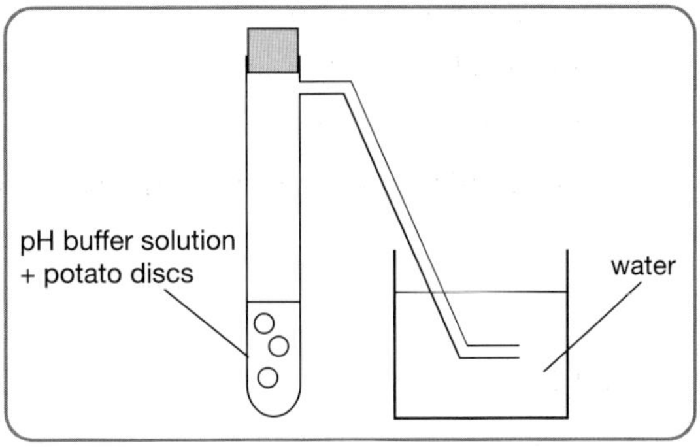

Figure 7.2 Apparatus for determining the factors that affect enzyme activity

Unit 3

Hess's law

Hess's law states that the overall energy change for a reaction is independent of the route taken.

The following reaction is studied.

$$KOH(s) + HCl(aq) \rightarrow KCl(aq) + H_2O(l)$$

This reaction can be carried out in one step, by adding solid KOH to HCl – or in two steps, by first dissolving solid KOH in water and then adding the solution of KOH to HCl.

The energy changes should be the same in each experiment.

One step

Solid KOH is added to 25 cm^3 of HCl solution.

The temperature is measured at the start and when the solid has dissolved.

The heat given out is calculated from $cm\Delta T$ and is scaled up to one mole of KOH.

Two steps

Solid KOH is added to 25 cm^3 of **water**.

The temperature is measured at the start and when the solid has dissolved. The heat given out is calculated and scaled up as above.

This solution is added to 25 cm^3 of HCl, and the temperature rise measured. The starting temperature is the average of the two solutions.

The heat given out is calculated and scaled up as above.

Quantitative electrolysis

The electrolysis of dilute sulphuric acid gives hydrogen at the negative electrode.

The equation is:

$$2H^+ + 2e^- \rightarrow H_2$$

The apparatus shown in Figure 7.3 is used.

The circuit includes an ammeter, a variable resistor and a power supply.

The measuring cylinder, at the start, should not be over the carbon rod.

The power is switched on, and the current adjusted to 0·5A.

The current runs for a few minutes to saturate the carbon rod with hydrogen.

The current is switched off and the measuring cylinder placed over the carbon rod.

The current is then switched on, and the time taken to collect around 50 cm^3 of gas is noted. The variable resistor is used to keep the current steady.

The current, time and volume of hydrogen are noted.

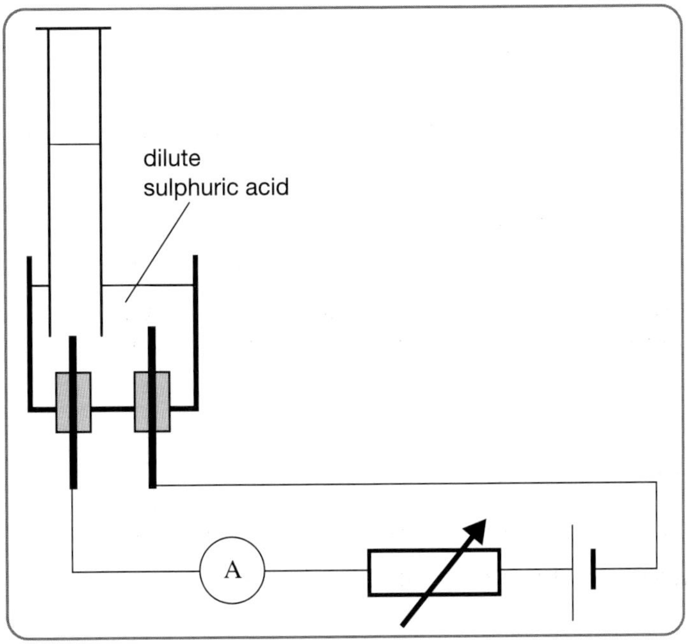

Figure 7.3 Apparatus for electrolysis

The number of coulombs is calculated from amps × seconds and hence the number of coulombs required to form one mole of gas (24 litres) can also be calculated.

The main sources of error are:

- current not constant
- error in measuring gas volume
- hydrogen slightly soluble in water.

Redox titration

The aim is to find the mass of vitamin C in a tablet, by carrying out a redox titration, in which Vitamin C reduces iodine to iodide ions.

The ion–electron equations are:

$$C_6H_8O_6 \text{ (Vitamin C)} \rightarrow C_6H_6O_6 + 2H^+ + 2e^-$$

$$I_2 + 2e^- \rightarrow 2I^-$$

Step 1: making a solution of vitamin C

A tablet of vitamin C is dissolved in 50 cm^3 of deionised water.

The solution is added to a graduated (standard) flask and the beaker rinsed into the flask several times with water.

Water is added carefully to the flask up to the mark on the neck of the flask.

The flask is inverted a few times to mix thoroughly.

Step 2: the titration

The pipette is rinsed out with vitamin C solution.

The burette is rinsed out with iodine solution.

25 cm^3 portions of vitamin C are transferred by pipette to conical flasks with starch indicator. Iodine solution is added from the burette until the colour changes to blue/black.

The titration is repeated to give at least two additional concordant results (agreeing to 0·1 cm^3).

The results are used to work out the mass of Vitamin C in the tablet.

Example 1

To determine the enthalpy of combustion of ethanol, a pupil heated 100 cm^3 of water using a spirit burner.

a) Draw a labelled diagram of the apparatus.

b) The experimental values of enthalpies of combustion are lower than data booklet values.

Give two sources of **experimental** error resulting in a lower enthalpy of combustion.

Solution

a) The diagram would be something like Figure 7.1 on page 114.

b) The main sources of experimental error are heat losses to the surroundings or incomplete combustion of the alcohol. The evidence for this is that the bottom of the copper container becomes quite sooty, as a result of carbon not being burned to carbon dioxide.

You might also be asked what measurements the pupil would have made. These are:

- starting and finishing temperature of the water
- starting and finishing mass of the burner
- mass of water.

Watch out – if you're asked what measurements are made, it's not correct to say 'temperature rise' or 'mass loss' because these are calculations, not measurements.

Example 2

The concentration of a solution of vitamin C can be found by adding standard iodine solution from a burette. The vitamin C reduces the iodine to iodide ions.

Starch is used as an indicator.

Example 2 *continued* ➢

Example 2 continued

The first steps are:

1. Add a vitamin C tablet to about 100 cm^3 of water in a beaker and stir to dissolve.
2. Transfer the solution quantitatively to a 500 cm^3 graduated flask.
   a) How would the solution be transferred quantitatively?
   b) What colour change occurs at the end point?

Solution

a) The solution would be poured carefully, using a funnel, into the graduated flask. The beaker would then be rinsed out several times with water, and the rinsings also transferred to the graduated flask. This guarantees that all the vitamin C is transferred.

b) In this reaction, iodine is added from the burette. When it meets the vitamin C, it is reduced to iodide. Since it becomes iodide, it has no effect on the starch. When all the vitamin C has reacted, there is nothing to turn iodine into iodide and so iodine builds up in the mixture. As a result, the mixture of starch and iodine changes colour to blue/black.

! Watch out when the question says '...what colour change...?' you have to give the starting colour and the finishing colour. If it started out as colourless, you need to say 'colourless to blue/black'.

Example 3

The ester ethyl propanoate can be made as follows from ethanol and propanoic acid.

a) What catalyst is added?

b) What is the purpose of the wet paper towel?

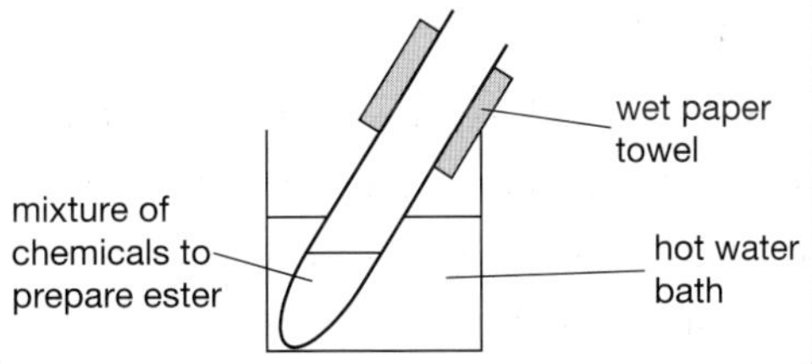

Figure 7.4 Apparatus for making an ester

Solution

a) The catalyst is concentrated sulphuric acid. You have to say that it's concentrated – if you don't, then you'll lose half a mark.

b) The ester is a lot more volatile than the chemicals it's made from, because unlike propanoic acid and ethanol it has no –OH groups and can't take part in hydrogen bonding. Being volatile, it turns readily into vapour. The wet paper towel is supposed to act as a condenser, and prevent the ester being lost into the atmosphere.

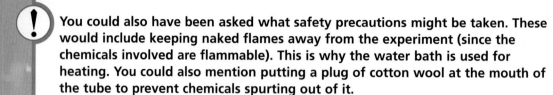

You could also have been asked what safety precautions might be taken. These would include keeping naked flames away from the experiment (since the chemicals involved are flammable). This is why the water bath is used for heating. You could also mention putting a plug of cotton wool at the mouth of the tube to prevent chemicals spurting out of it.

You could also be asked how the ester would be separated from the mixture. In the PPA, you do this by pouring the mixture into a small beaker containing a solution of sodium hydrogen carbonate. The catalyst is neutralised by this, so can't catalyse the breakdown of the ester, and the ester floats on the surface, as a separate layer.

Example 4

The effect of temperature change on rate of reaction can be studied using the reaction between acidified potassium permanganate solution and a dicarboxylic acid with the formula $(COOH)_2$.

The equation for the reaction is:

$$5(COOH)_2 + 6H^+ + 2MnO_4^- \rightarrow 2Mn^{2+} + 10CO_2 + 8H_2O$$

a) Name the dicarboxylic acid.

b) Describe how the reaction time is measured.

c) Why is it difficult to measure accurately the reaction time at room temperature?

Solution

a) Doh! You're supposed to know this! It's oxalic acid.

b) In this reaction, the purple permanganate ions end up as the almost colourless Mn^{2+} ions. The mixture gradually turns colourless. You measure the time taken for this to happen.

c) The reaction is very slow so that the colour change is too gradual.

LAST WORD

On the Day – Advice and Reminders

Preferably before the exam, make sure you know the date, time and place of the exam! Arrive 15 minutes early and bring with you two pens, a pencil and sharpener, a ruler and calculator – as we said before, one that you know how to use, in working order and with fresh batteries. Don't bring your mobile phone. A Data Booklet will be provided.

Tackle the multiple choice part (Section A) first. Indicate an answer, using a horizontal line in ink. Make sure you note, on page 2 of the exam booklet, how to make changes to multiple choice answers. Don't leave out any questions.

Read the questions carefully and make sure you're taking in what the question is asking. Be very careful with questions that contain the words 'both' and 'neither'/'increase' and 'decrease'/'greatest' and 'least' to make sure that you really understand what the question is asking.

In Section B, again it's a good idea to read the questions. Start with the first question and work through the questions. If you meet one you can't do, skip it, and return to it later. But make sure that you **do** return.

In calculations, show as much working as you can. Try to let the marker see what thought processes you've been using. Be careful with the units of your answer – check the wording to see if they are given in the question, and, if so, it's not necessary to give them. Make sure that your answers are reasonable – for example, if you're asked to work out the number of atoms or molecules in a certain mass of an element or compound, the answer is bound to be a very large number. If you're asked to work out the mass of an atom or molecule, it's going to be a very small number.

If you're asked to draw or complete a diagram, use a pencil and ruler. Try to make it neat. Use real, recognisable apparatus in your drawing. You're best to stick to cross-sectional type diagrams, rather than try an artistic three dimensional drawing. Cross-sectional diagrams are easier to do, and less confusing to interpret. Make sure you don't close any tubes which are meant to be open.

Don't be put off by questions which seem totally unfamiliar to you (unless you've done no revision, in which case they will all be totally unfamiliar!). Your paper will contain problem solving questions (often with lots of words) involving reactions and techniques you haven't seen before. Take time, read the question a couple of times, and try to think it through.

Good luck!